Methods and Protocols in Food Science

Series Editor
Anderson S. Sant'Ana
University of Campinas
Campinas, Brazil

For further volumes:
http://www.springer.com/series/16556

Methods and Protocols in Food Science series is devoted to the publication of research protocols and methodologies in all fields of food science.

Volumes and chapters will be organized by field and presented in such way that the readers will be able to reproduce the experiments in a step-by-step style. Each protocol will be characterized by a brief introductory section, followed by a short aims section, in which the precise purpose of the protocol will be clarified.

Meat Product Quality Improvement

Strategies and Technology

Edited by

José Manuel Lorenzo Rodriguez

Centro Tecnolóxico da Carne de Galicia, Parque Tecnológico de Galicia, San Cibrao das Viñas, Ourense, Spain; Universidade de Vigo, Área de Tecnología de los Alimentos, Facultad de Ciencias de Ourense, Ourense, Spain

Humana Press

Editor
José Manuel Lorenzo Rodriguez
Centro Tecnolóxico da Carne de Galicia
Parque Tecnológico de Galicia
San Cibrao das Viñas, Ourense, Spain

Universidade de Vigo
Área de Tecnología de los Alimentos
Facultad de Ciencias de Ourense
Ourense, Spain

ISSN 2662-950X ISSN 2662-9518 (electronic)
Methods and Protocols in Food Science
ISBN 978-1-0716-5459-0 ISBN 978-1-0716-5460-6 (eBook)
https://doi.org/10.1007/978-1-0716-5460-6

This Humana imprint is published by the registered company Springer Science+Business Media, LLC, part of Springer Nature.
The registered company address is: 1 New York Plaza, New York, NY 10004, U.S.A.

Preface to the Series

Methods and Protocols in Food Science series is devoted to the publication of research protocols and methodologies in all fields of food science. The series is unique as it includes protocols developed, validated and used by food and related scientists as well as theoretical basis are provided for each protocol. Aspects related to improvements in the protocols, adaptations and further developments in the protocols may also be approached.

Methods and Protocols in Food Science series aims to bring the most recent developments in research protocols in the field as well as very well established methods. As such the series targets undergraduate, graduate and researchers in the field of food science and correlated areas. The protocols documented in the series will be highly useful for scientific inquiries in the field of food sciences, presented in such way that the readers will be able to reproduce the experiments in a step-by-step style.

Each protocol will be characterized by a brief introductory section, followed by a short aims section, in which the precise purpose of the protocol is clarified. Then, an in-depth list of materials and reagents required for employing the protocol is presented, followed by a comprehensive and step-by-step procedures on how to perform that experiment. The next section brings the do's and don'ts when carrying out the protocol, followed by the main pitfalls faced and how to troubleshoot them. Finally, template results will be presented and their meaning/conclusions addressed.

The *Methods and Protocols in Food Science* series will fill an important gap, addressing a common complain of food scientists, regarding the difficulties in repeating experiments detailed in scientific papers. With this, the series has a potential to become a reference material in food science laboratories of research centers and Universities throughout the world.

Campinas, Brazil ***Anderson S. Sant'Ana***

Preface

Consumers are increasingly aware of the impact of diet on health and are well-informed about these issues. As a result, they demand healthier meat products and are also willing to pay an additional cost for them. In this regard, the elaboration of meat products with modified composition (reformulation) is a central topic of research, development, and innovation in the meat industry. The modifications are motivated by many factors related to health, lifestyle, and consumer values and motives at the time of purchase. However, the reformulation of meat products is a complex process that can lead to important changes in their composition and, therefore, can affect the characteristics associated with this type of product. This condition creates the need to study and develop strategies that allow to achieve a suitable match between the expectations and needs of the consumer and the characteristics and attributes of the meat.

This book focuses on exploring strategies to improve meat quality using trending topics in Food Quality and Safety. The aim of the present book is to cover the research works carried out in favor of healthy meat products and collect and systematize the existing knowledge in this field. Specific topics covered by this book include (a) the reduction of salt content in meat products; (b) the implementation of new technologies to reduce the use of additives in meat products; (c) strategies to reduce fat content in meat products; (d) the inclusion of natural colorants to extend the shelf-life of meat products; and (e) the use of probiotics, prebiotics, and postbiotics in meat products. Although the "Material and Methods" section is imperative in any scientific article about meat products, the number of comprehensive publications with detailed information on the applied methods is limited. This book will fulfill the current scarcity of detailed and comprehensive sources of scientific information about the methods applied in the study of meat products.

The book covers the main strategies to obtain healthy meat products in 17 chapters, including the techniques, processing conditions, and ingredients used in the manufacture of meat products. A complete and comprehensive description of the materials and processing conditions used is made so that each meat product can be manufactured by other researchers or industries. In addition, each chapter includes explanatory notes and highlights the specific points to consider for correct manufacturing. This book explores diverse technological strategies to develop healthier meat alternatives without compromising traditional sensory attributes. A primary focus is the improvement of lipid profiles, achieved by replacing animal fats with structured systems such as gelled emulsions, oleogels, and microencapsulated vegetable oils rich in n-3 PUFAs. Simultaneously, significant efforts are being made to reduce sodium levels in dry-cured products like cecina and rabbit leg through the partial replacement of NaCl with alternative minerals like potassium chloride or kombu seaweed.

Furthermore, the valorization of agro-industrial co-products—such as peels from kiwi, prickly pear, elderberry, xoconostle, and purple sweet potato—serves a dual purpose by providing natural antioxidants and colorants while promoting a circular economy. The integration of emerging technologies, including high-power ultrasound for phosphate-free emulsions and spray-drying for the encapsulation of bioactive compounds, ensures the stability and efficacy of these natural ingredients throughout the manufacturing process. The incorporation of probiotics, such as *Lactiplantibacillus plantarum*, into dry-fermented sausages like salamines criollos and chorizo further elevates these products into the category of functional foods, offering health benefits that extend beyond basic nutrition. By providing detailed, replicable protocols for these innovations, this work aims to support both the meat industry and the scientific community in creating high-quality, sustainable, and health-conscious meat products that satisfy modern consumer expectations.

San Cibrao das Viñas, Ourense, Spain *José Manuel Lorenzo Rodriguez*

Contents

Contributors

Rubén Agregán • *Centro Tecnolóxico da Carne, Parque Tecnolóxico de Galicia, San Cibrao das Viñas, Ourense, Spain*

Esteban Arteaga-Cabrera • *Posgrado en Biociencias, División de Ciencias de la Vida, Campus Irapuato-Salamanca Universidad de Guanajuato, Irapuato, Guanajuato, México*

Juliano Smanioto Barin • *Universidade Federal de Santa Maria, Santa Maria, Rio Grande do Sul, Brazil*

Julliane Carvalho Barros • *Universidade de Sao Paulo Faculdade de Zootecnia e Engenharia de Alimentos, Department of Food Engineering, Pirassununga, Brazil*

Carmen Botella-Martínez • *IPOA Research Group, Institute for Agri-Food and Agri-Environmental Research and Innovation (CIAGRO-UMH), Miguel Hernández University, Orihuela, Alicante, Spain*

Paulo Cezar Bastianello Campagnol • *Universidade Federal de Santa Maria, Santa Maria, Rio Grande do Sul, Brazil*

Daliana Carbajal-Padilla • *Posgrado en Biociencias, División de Ciencias de la Vida, Campus Irapuato-Salamanca Universidad de Guanajuato, Irapuato, Guanajuato, México*

Carlos Pasqualin Cavalheiro • *INDMEAT Group (Innovation in the Development, Characterization and Storage of Meat Products), Department of Meat and Fish Products, Institute of Food Science, Technology and Nutrition (ICTAN-CSIC), Madrid, Spain; Laboratório de Inspeção e Tecnologia de Carnes e Derivados (LabCarne), Universidade Federal da Bahia (UFBA), Salvador, Brazil*

Magdalena I. Ceron Guevara • *Universidad Autónoma del Estado de Hidalgo, Mineral de la Reforma, Hidalgo, Mexico*

Alexandre José Cichoski • *Universidade Federal de Santa Maria, Santa Maria, Rio Grande do Sul, Brazil*

Aurora Cittadini • *Centro Tecnolóxico da Carne, Parque Tecnolóxico de Galicia, San Cibrao das Viñas, Ourense, Spain; Centro Tecnolóxico da Carne, Ourense, Spain*

Rubén Domínguez-Valencia • *Centro Tecnolóxico da Carne de Galicia, Parque Tecnológico de Galicia, San Cibrao das Viñas, Ourense, Spain*

Noemí Echegaray • *Centro Tecnolóxico da Carne, Parque Tecnolóxico de Galicia, San Cibrao das Viñas, Ourense, Spain*

Juana Fernández-López • *Research Group in Innovations in Food Products (IPOA), Institute for Agri-Food and Agri-Environmental Research and Innovation, Miguel Hernández University (CIAGRO-UMH), Orihuela, Alicante, Spain*

Laureano Sebastián Frizzo • *Laboratory of Food Analysis "Med. Vet R. Dalla Santina", Institute of Veterinary Science (ICiVet Litoral), National University of the Littoral (UNL)-National Scientific and Technical Research Council (CONICET), Esperanza, Santa Fe, Argentina; Department of Public Health, Faculty of Veterinary Sciences, National University of the Littoral (UNL), Esperanza, Santa Fe, Argentina*

JULIÁN A. GÓMEZ-SALAZAR • *Posgrado en Biociencias, División de Ciencias de la Vida, Campus Irapuato-Salamanca Universidad de Guanajuato, Irapuato, Guanajuato, México*
ROSANE TERESINHA HECK • *Universidade Federal de Santa Maria, Santa Maria, Rio Grande do Sul, Brazil*
ANA MARIA HERRERO • *INDMEAT Group (Innovation in the Development, Characterization and Storage of Meat Products), Department of Meat and Fish Products, Institute of Food Science, Technology and Nutrition (ICTAN-CSIC), Madrid, Spain*
FLORENCIA AYLEN LENCINA • *Laboratory of Food Analysis "Med. Vet R. Dalla Santina", Institute of Veterinary Science (ICiVet Litoral), National University of the Littoral (UNL)-National Scientific and Technical Research Council (CONICET), Esperanza, Santa Fe, Argentina*
JOSÉ MANUEL LORENZO RODRIGUEZ • *Centro Tecnolóxico da Carne de Galicia, Parque Tecnológico de Galicia, San Cibrao das Viñas, Ourense, Spain; Universidade de Vigo, Área de Tecnología de los Alimentos, Facultad de Ciencias de Ourense, Ourense, Spain*
RAQUEL LUCAS-GONZÁLEZ • *Institute for Agri-Food and Agri-Environmental Research and Innovation, Miguel Hernández University (CIAGRO-UMH), Alicante, Spain*
DANUBIA PERPÉTUA MAIOTO DE SOUZA • *São Paulo State University – Unesp, Institute of Biosciences, Humanities and Exact Sciences–Ibilce, Department of Food Engineering and Technology, Campus São José Do Rio Preto, São Paulo, Brazil*
CRISTIANO RAGAGNIN DE MENEZES • *Universidade Federal de Santa Maria, Santa Maria, Rio Grande do Sul, Brazil*
GABRIELA XAVIER OIOLI • *São Paulo State University – Unesp, Institute of Biosciences, Humanities and Exact Sciences–Ibilce, Department of Food Engineering and Technology, Campus São José Do Rio Preto, São Paulo, Brazil*
MIRIAM PATEIRO • *Centro Tecnológico de la Carne de Galicia, Parque Tecnológico de Galicia, San Cibrao das Viñas, Ourense, Spain*
MIRIAN PATEIRO • *Centro Tecnolóxico da Carne, Parque Tecnolóxico de Galicia, San Cibrao das Viñas, Ourense, Spain; Área de Tecnología de los Alimentos, Facultad de Ciencias de Ourense, Universidad de Vigo, Ourense, Spain; Centro Tecnolóxico da Carne, Ourense, Spain*
DOUGLAS PEDRO • *Universidade Federal de Santa Maria, Santa Maria, Rio Grande do Sul, Brazil; Instituto Federal Farroupilha, Júlio de Castillhos, Rio Grande do Sul, Brazil*
JOSÉ ÁNGEL PÉREZ-ÁLVAREZ • *Institute for Agri-Food and Agri-Environmental Research and Innovation, Miguel Hernández University (CIAGRO-UMH), Orihuela, Alicante, Spain; IPOA Research Group, Institute for Agri-Food and Agri-Environmental Research and Innovation (CIAGRO-UMH), Miguel Hernández University, Orihuela, Alicante, Spain*
ALONSO PÉREZ-FERNÁNDEZ • *Institute for Agri-Food and Agri-Environmental Research and Innovation, Miguel Hernández University (CIAGRO-UMH), Orihuela, Alicante, Spain; Servicio de Farmacia, Hospital Universitario de La Princesa, Madrid, Spain*
MARIANA BASSO PINTON • *Universidade Federal de Santa Maria, Santa Maria, Rio Grande do Sul, Brazil*
ESMERALDA RANGEL-VARGAS • *Universidad Autónoma del Estado de Hidalgo, Mineral de la Reforma, Hidalgo, Mexico*
CLAUDIA RUIZ-CAPILLAS • *INDMEAT Group (Innovation in the Development, Characterization and Storage of Meat Products), Department of Meat and Fish Products, Institute of Food Science, Technology and Nutrition (ICTAN-CSIC), Madrid, Spain*
BIBIANA ALVES DOS SANTOS • *Universidade Federal de Santa Maria, Santa Maria, Rio Grande do Sul, Brazil*

EVA M. SANTOS • *Universidad Autónoma del Estado de Hidalgo, Mineral de la Reforma, Hidalgo, Mexico*
MARCELO LISANDRO SIGNORINI • *Laboratory of Food Analysis "Med. Vet R. Dalla Santina", Institute of Veterinary Science (ICiVet Litoral), National University of the Littoral (UNL)-National Scientific and Technical Research Council (CONICET), Esperanza, Santa Fe, Argentina; Epidemiology and Infectious Diseases, Dairy Chain Research Institute (IdICAL INTA-CONICET), National Institute of Agricultural Technology, Rafaela, Santa Fe, Argentina*
SUELEM LIMA DA SILVA • *Universidade Federal de Santa Maria, CEP, Santa Maria, Rio Grande do Sul, Brazil*
ANDREA CARLA DA SILVA BARRETTO • *São Paulo State University – Unesp, Institute of Biosciences, Humanities and Exact Sciences–Ibilce, Department of Food Engineering and Technology, Campus São José Do Rio Preto, São Paulo, Brazil*
NOELÍ ESTEFANÍA SIRINI • *Laboratory of Food Analysis "Med. Vet R. Dalla Santina", Institute of Veterinary Science (ICiVet Litoral), National University of the Littoral (UNL)-National Scientific and Technical Research Council (CONICET), Esperanza, Santa Fe, Argentina; Department of Public Health, Faculty of Veterinary Sciences, National University of the Littoral (UNL), Esperanza, Santa Fe, Argentina*
LORENA PAOLA SOTO • *Laboratory of Food Analysis "Med. Vet R. Dalla Santina", Institute of Veterinary Science (ICiVet Litoral), National University of the Littoral (UNL)-National Scientific and Technical Research Council (CONICET), Esperanza, Santa Fe, Argentina; Department of Public Health, Faculty of Veterinary Sciences, National University of the Littoral (UNL), Esperanza, Santa Fe, Argentina*
MARÍA ÁNGELES STEGMAYER • *Laboratory of Food Analysis "Med. Vet R. Dalla Santina", Institute of Veterinary Science (ICiVet Litoral), National University of the Littoral (UNL)-National Scientific and Technical Research Council (CONICET), Esperanza, Santa Fe, Argentina; Department of Public Health, Faculty of Veterinary Sciences, National University of the Littoral (UNL), Esperanza, Santa Fe, Argentina*
ALFREDO TEIXEIRA • *Mountain Research Centre (CIMO), Escola Superior Agrária, Instituto Politécnic o de Bragança, Bragança, Bragana, Portugal*
MARCO ANTONIO TRINDADE • *Universidade de Sao Paulo Faculdade de Zootecnia e Engenharia de Alimentos, Department of Food Engineering, Pirassununga, Brazil*
MANUEL VIUDA-MARTOS • *Research Group in Innovations in Food Products (IPOA), Institute for Agri-Food and Agri-Environmental Research and Innovation, Miguel Hernández University (CIAGRO-UMH), Orihuela, Alicante, Spain*
ROGER WAGNER • *Universidade Federal de Santa Maria, Santa Maria, Rio Grande do Sul, Brazil*
MARÍA VIRGINIA ZBRUN • *Laboratory of Food Analysis "Med. Vet R. Dalla Santina", Institute of Veterinary Science (ICiVet Litoral), National University of the Littoral (UNL)-National Scientific and Technical Research Council (CONICET), Esperanza, Santa Fe, Argentina; Epidemiology and Infectious Diseases, Dairy Chain Research Institute (IdICAL INTA-CONICET), National Institute of Agricultural Technology, Rafaela, Santa Fe, Argentina*
JORGE ALBERTO ZIMMERMANN • *Laboratory of Food Analysis "Med. Vet R. Dalla Santina", Institute of Veterinary Science (ICiVet Litoral), National University of the Littoral (UNL)-National Scientific and Technical Research Council (CONICET), Esperanza, Santa Fe, Argentina; Physiology, Department of Preclinical Sciences, Faculty of Veterinary Sciences, National University of the Littoral (UNL), Esperanza, Santa Fe, Argentina*

Chapter 1

Production of Burgers Rich in n-3 PUFAs Using Microencapsulated Chia and Linseed Oils

Rosane Teresinha Heck, Jose Manuel Lorenzo Rodriguez, Alexandre José Cichoski, Cristiano Ragagnin de Menezes, Juliano Smanioto Barin, Roger Wagner and Paulo Cezar Bastianello Campagnol

Abstract

This chapter provides a step-by-step protocol for producing low-fat burgers with improved lipid profiles by replacing 50% pork back fat with microencapsulated chia or linseed oils. The oils are encapsulated using external ionic gelation to form thermally stable microparticles, which are incorporated into the meat matrix without impairing technological or sensory quality. The chapter details the preparation of the microparticles, burger formulation, and processing and includes practical recommendations for achieving healthier burgers rich in omega-3 polyunsaturated fatty acids (PUFAs).

Keywords Low-fat burgers, Microencapsulation, Chia oil, Linseed oil, Omega-3, PUFA, Healthier meat products

1 Introduction

Traditional burgers are often high in fat, particularly saturated fatty acids (SFAs), and display an unbalanced n−6/n−3 PUFA ratio, both of which have been associated with increased risk of chronic diseases, such as cardiovascular disorders and certain cancers [1]. Consequently, there is a growing demand for healthier meat products with reduced animal fat and improved nutritional profiles.

Chia (*Salvia hispanica L.*) and linseed (*Linum usitatissimum*) oils are rich in n−3 PUFAs, especially alpha-linolenic acid (ALA), which has been shown to reduce cardiovascular risk [2]. However, directly incorporating these oils into meat products may compromise texture, oxidative stability, and shelf life [3]. Microencapsulation using external ionic gelation with alginate and calcium ions provides a protective matrix that enhances oil stability during processing and storage [4].

José Manuel Lorenzo Rodriguez (ed.), *Meat Product Quality Improvement*, Methods and Protocols in Food Science,
https://doi.org/10.1007/978-1-0716-5460-6_1,

This chapter presents a comprehensive protocol for incorporating microencapsulated chia or linseed oils into burgers by replacing 50% of pork back fat usually added in this product. The process ensures improved PUFA/SFA and n−6/n−3 ratios without negatively affecting the product's technological and sensory properties.

2 Materials

Ingredients:

- Beef (Rectus femoris; 78.4%)
- Pork back fat (10.0%)
- Microencapsulated chia or linseed oil (10.0%)
- Sodium chloride (1.5%)
- Garlic powder (0.1%)

Microparticles preparation:

- Chia or linseed oil (25%)
- Sodium alginate solution (2.0%)
- Calcium chloride solution (0.1 M)
- Sterile distilled water

Equipment:

- Dual-fluid atomizer nozzle (0.1 mm)
- Compressed air source (0.125 kg/cm^2)
- Magnetic stirrer
- 150 μm wire mesh sieve
- Meat grinder with 3 mm disc
- Bowl mixer
- Burger forming machine
- Freezer (−18°C)

3 Methods

3.1 Preparation of Microparticles

1. Prepare a 2.0% sodium alginate solution in distilled water. Stir continuously with a magnetic stirrer for at least 2 h until fully hydrated (*see* **Note 1**).
2. Mix chia or linseed oil into the alginate solution to form a stable emulsion. The final oil content should represent 25% of the total emulsion volume (*see* **Note 2**).

3. Using a dual-fluid nozzle (0.1 mm), atomize the emulsion into a gently stirred 0.1 M $CaCl_2$ solution. Maintain the nozzle 12 cm above the solution surface and regulate air pressure to 0.125 kg/cm² (*see* **Note 3**).
4. Allow the resulting microparticles to harden in the $CaCl_2$ solution for 30 min while stirring slowly.
5. Filter microparticles through a 150 μm sieve and rinse gently with sterile distilled water to remove residual calcium (Fig. 1a) (*see* **Note 4**). Figure 1 shows the production of microparticles (a), Microencapsulated chia (T1) and linseed (T2) oils (b).
6. Store the microparticles at 4°C in sealed containers and use within 48 h (*see* **Note 5**).

3.2 Burger Formulation and Production

1. Grind beef and pork back fat separately using a 3 mm plate.
2. In a bowl mixer, combine ground beef with sodium chloride and mix until protein extraction is evident (approximately 2–3 min).
3. Add garlic powder, pork back fat, and microencapsulated chia or linseed oil (Fig. 1b) (10% of total weight).
4. Mix gently until a homogeneous batter is obtained (*see* **Note 6**).
5. Shape into 100 g burger patties (11 cm diameter, 2.5 cm height) using a forming machine.

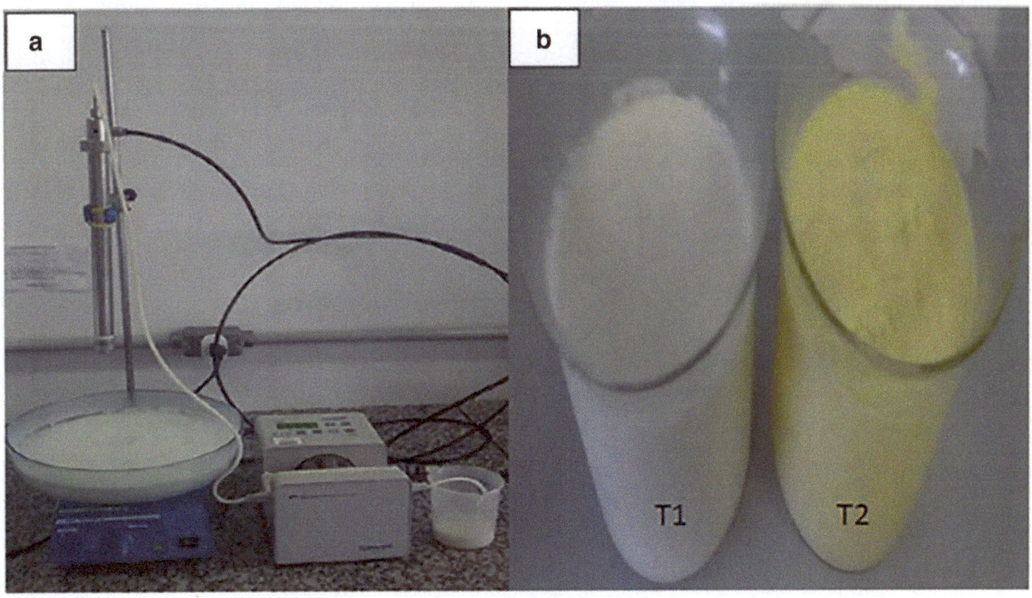

Fig. 1 **(a)** Production of microparticles. **(b)** Microencapsulated chia (T1) and linseed (T2) oils

6. Freeze immediately at −18°C to preserve structure and avoid oil leakage (*see* **Note 7**).

4 Notes

1. Complete hydration of sodium alginate is essential for uniform emulsion formation and efficient encapsulation.
2. To avoid phase separation, ensure slow and uniform mixing when incorporating oil into alginate.
3. To ensure consistent capsule size, the atomization step must be performed under constant air pressure and fixed distance.
4. Do not use excessive force during sieving and washing to prevent structural damage to the microparticles.
5. Microparticles should be used within 48 h of preparation and refrigerated at 4°C to maintain structural integrity.
6. Avoid overmixing to minimize rupture of the microcapsules and ensure uniform distribution throughout the meat matrix.
7. Freezing should be performed rapidly to preserve the capsules' mechanical integrity and the lipids' stability.

References

1. World Health Organization (2003) Diet, nutrition and the prevention of chronic diseases: report of a joint WHO/FAO expert consultation, Geneva, 28 January–1 February 2002. WHO Technical Report Series, No. 916. World Health Organization, Geneva. ISBN 92-4-120916-X
2. Ayerza R, Coates W (2005) Ground chia seed and chia oil effects on plasma lipids and fatty acids in the rat. Nutr Res 25(11):995–1003. https://doi.org/10.1016/j.nutres.2005.09.013
3. Heck RT, Dos Santos BA, Lorenzo JM et al (2022) Replacement of saturated fat by healthy oils to improve nutritional quality of meat products. Food Lipids 2022:461–487. https://doi.org/10.1016/B978-0-12-823371-9.00008-3
4. Heck RT, Lorenzo JM, Dos Santos BA et al (2021) Microencapsulation of healthier oils: an efficient strategy to improve the lipid profile of meat products. Curr Opin Food Sci 40:6–12. https://doi.org/10.1016/J.COFS.2020.04.010

Chapter 2

Alheiras with Improved Lipid Profile Using Gelled Emulsions with Vegetable Oils

Juana Fernández-López, Carmen Botella-Martínez, Alfredo Teixeira, Manuel Viuda-Martos and José Ángel Pérez-Álvarez

Abstract

Alheira is one of the most traditional processed pork products from *Trás-os-Montes*, northeastern Portugal, and is widely consumed across the country as part of Portuguese gastronomy. Nowadays, its popularity is decreasing because of its high saturated fat content and the negative consumer perception related to their health consequences. Therefore, trying to improve the lipid profile of alheiras has become a major challenge for the meat industry. This study describes the formula, processing stages, chemical composition, and lipid profile of alheiras, focusing on its reformulation to enhance its lipid profile. Partial substitution of fat raw materials by gelled emulsion made with vegetables oils has resulted a technologically feasible strategy to improve the alheiras lipid profile while maintaining their characteristic properties.

Keywords Alheira, Sausage, Fat replacement, Gelled emulsion, Vegetable oil

1 Introduction

Alheira is a traditional sausage from Portugal, primarily made from poultry, bread, olive oil, lard, garlic, and paprika. Originally known as "the sausage of the Jews," it was shaped like a horseshoe and crafted by Portuguese Jews during the Inquisition in the late 16th and 17th centuries. These individuals, who identified as "new Christians," secretly adhered to their Jewish customs, which included avoiding pork [1]. As a result, the early versions of alheira featured various types of meat, such as turkey and chicken, as substitutes for pork. Over time, the sausage gained popularity and now often includes a mix of meats, particularly fatty pork. While it remains a traditional artisanal product, the demand has led to industrial production, resulting in many alheiras available (Fig. 1).

Since its recognition as a Traditional Specialty Guaranteed (TSG) product in 1996, alheira has also been granted Protected Geographical Indication (PGI) status by the European Union for certain

José Manuel Lorenzo Rodriguez (ed.), *Meat Product Quality Improvement*, Methods and Protocols in Food Science,
https://doi.org/10.1007/978-1-0716-5460-6_2,

Fig. 1 Industrial production of alheiras (drying, packing, and commercialization)

regional varieties [2]. Nevertheless, numerous commercial versions lacking PGI certification are produced across Portugal, particularly in areas outside the Northeast region, which is the source of alheira.

Despite its widespread popularity, consumer perception of alheira is often negative due to its high fat content, which has become a significant concern within the meat industry. In response, the sector is actively pursuing various strategies aimed at reducing or eliminating saturated fats from product formulations. The objective is to develop high-quality, nutritious, and healthier meat alternatives while preserving traditional flavors and satisfying consumer expectations. To achieve this, a range of methods has been employed to substitute animal-derived saturated fats with vegetable oils, thereby minimizing their effects on the physical, chemical, and sensory characteristics of the final products. This approach ensures consumer acceptance, adherence to essential technological standards, and economic viability within the meat industry [3–9]. Among these strategies, the utilization of gelled emulsions (GE) has emerged as a promising technique for structuring liquid oils derived from diverse sources, including marine, vegetable, and seed oils, which are characterized by healthier fatty acid profiles and aim to mimic the functional properties of pork backfat [10, 11].

Furthermore, various emulsifiers, such as chestnut flour, chia mucilage flour, and soy protein, as well as gelling agents such as gelatin, alginate, carrageenan, and different gums, have been employed in the formulation of these gelled emulsions [3, 4, 6, 12, 13]. As a result, gelled emulsions incorporating various vegetable oils and gelling agents have been successfully utilized as fat substitutes in the development of both cooked and fresh meat products, leading to enhanced health benefits [5, 8, 13–15].

The aim of these studies is to enhance the understanding of this well-loved product in terms of its characteristics and nutritional quality. Consequently, this chapter will provide a comprehensive overview of the manufacturing process and formulation of alheira, with a focus on improving its lipid profile by incorporating gelled emulsions as fat substitute.

2 Materials

According to Commission Regulations (EC) [16–18], alheira is a meat product made from a blend of pork and poultry meat, with the option to include other types of poultry or game, such as partridge (*see* **Note 1**). The meats are first boiled, then deboned, chopped, or shredded, and combined with finely sliced bread and other ingredients such as garlic, olive oil, and paprika. The resulting mixture is uneven, forming a soft and slightly lumpy bread pulp infused with broth and pieces of shredded meat. The broth used for cooking the meat further softens the mixture (*see* **Note 2**). As a source of fat for the production of the alheiras, pork belly or a far substitute, a gelled emulsion based on hemp oil and amaranth flour was used according to the study carried out by [19]. The dough is then encased in natural casings (derived from pork or cattle), lightly smoked, dried, and left to mature for at least 8 days. While considered a cooked sausage, it is best consumed cooked after the maturation process, either grilled or deep-fried. Its sensory attributes include:

1. Weight and dimensions: the weight will be variable, ranging from a minimum of 150 g to a maximum of 200 g. It measures approximately 25–30 cm in length with a diameter of 2 cm.
2. Shape and external appearance resemble a horseshoe.
3. Color: bright yellow-brownish tone. After undergoing heat treatment, it transitions to a reddish, tanned brown.
4. Aroma: pleasant.
5. Taste: pleasant taste, with olive and garlic tastes.
6. Texture: uneven texture, characterized by noticeable pieces of shredded meat.

3 Methods

The various stages of the manufacturing process will be conducted in adherence to the traditional expertise of the producers, ensuring the production of high-quality products. The elaboration process of alheira included the selection of the meat, cooking, mixing of ingredients, salting, cutting and shredding, and curing, which will be carried out as described below (Fig. 2). The gelled emulsion is also prepared beforehand, and 25% and 50% of pork belly used is subsequently replaced by the gelled emulsion [19] (Fig. 3).

1. The selection of the meat (duck, chicken, pork belly, and partridge) is carried out at a maximum of 6°C of temperature.
2. Cook all meats for 5 h above 95°C in salty water (2% of salt) (*see* **Note 3**).
3. Cut bread into thin slices (*see* **Note 4**).
4. Mixed the bread with the shredded meat (*see* **Note 5**).
5. Pour in a quantity of the cooking broth into the mixture.
6. Mixed the other ingredients: paprika, garlic, and olive oil.
7. Mix with the fat, either the pork belly, the gelled emulsion or both.
8. Filling in natural casings and horseshoe shape.
9. Stabilization for 90 min in an oven from 23°C to 71°C.
10. Heat treatment in an oven at 60°C for 6 h.
11. Drying at 30°C from humidity from 85% to 65%–70%.

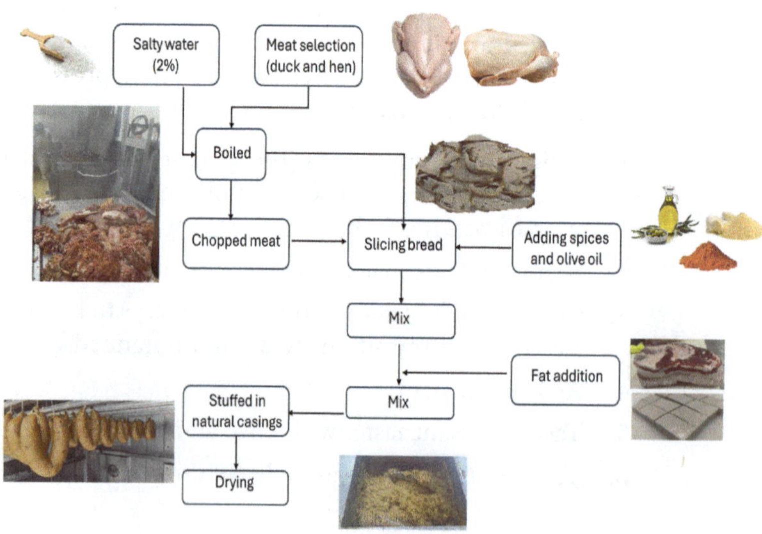

Fig. 2 Diagram of the production process of alheira with improved lipid profile

Fig. 3 Different stages of the production process of alheira with improved lipid profile

The chemical composition of the final products is presented in Table 1. Significant differences can be observed between the control sample (traditional alheira) and the reformulated alheiras (AGEH25 and AGEH50) in which the animal fat was replaced by the gelled emulsion (with hemp oil): AGEH25- with a 25% animal fat substitution and AGEH50- with a 50% animal fat substitution. Alheiras with higher protein content (2.6%–6% more protein) and lower fat content (7.5%–23% less fat) than control Alheiras were successfully obtained. In addition, reformulated Alheiras showed lower saturated fatty acids content (5.9%–16.4% less) and higher polyunsaturated fatty acid content (33%–87% more) than traditional Alheiras.

4 Notes

1. The typical meat blend used in alheira consists of pork and poultry; however, it may also include other types of meat, such as rabbit, duck, turkey, or game meats like partridge. The original version known as “alheira of the Jews” did not contain any pork.
2. The alheira that carries the PGI designation, along with the bread and olive oil, is traditional products originating from the *Trás-os-Montes* region in northeastern Portugal.

Table 1
Proximate composition (g/100 g) and main fatty acids (g/100 g of fat) of alheiras with improved lipid profile using gelled emulsions with vegetable oils

	Control	AGEH25	AGEH50
Moisture	56.52	58.03	59.87
Protein	9.85	11.63	12.18
Fat	15.07	13.94	11.63
Ash	1.17	1.35	1.38
Chlorides	0.78	0.89	0.91
pH	5.95	5.79	5.92
SFA	36.95	34.75	30.90
MUFA	48.18	45.50	41.83
PUFA	14.86	19.76	27.27
n3	0.54	1.87	3.68
n6	14.30	17.85	23.56

SFA saturated fatty acids, *MUFA* monounsaturated fatty acids, *PUFA* polyunsaturated fatty acids, *n3* omega 3 fatty acids, *n6* omega 6 fatty acids. Data obtained by published article [19]

3. The cooking process should be conducted at an intensity that facilitates the deboning of the meat.
4. A traditional regional bread, weighing between 2 and 3 kg, is made from wheat and is baked with minimal yeast.
5. The bread, which is sliced finely, must be soaked in the meat cooking broth at a temperature ranging from 70°C to 80°C before being combined with the meat.

Acknowledgments

This work was supported by CIAGRO-UMH. Thanks to Generalitat Valenciana for supporting this book chapter through the AICO 2024 grant for Consolidated Research Groups (grant number CIAICO/2023/004). The authors are members of the Healthy Meat network, funded by CYTED (ref. 119RT0568).

References

1. Teixeira A, Fernandes A, Pereira E (2020) Fat content reduction and lipid profile improvement in Portuguese fermented sausages alheira. Heliyon 6:e05306. https://doi.org/10.1016/j.heliyon.2020.e05306
2. Teixeira A, Rodrigues S (2019) Meat quality, brands and consumer trends. In: Lorenzo J, Munekata P, Barba F, Toldrá F (eds) More than beef, pork and chicken—the production, processing, and quality traits of other sources of meat for human diet. Springer, Cham. https://doi.org/10.1007/978-3-030-05484-7-211
3. Lucas-González R, Roldán-Verdu A, Sayas-Barberá E et al (2020) Assessment of emulsion gels formulated with chestnut (*Castanea sativa* M.) four and chia (*Salvia hispanica* L) oil as partial fat replacers in pork burger formulation. J Sci Food Agric 100:1265–1273. https://doi.org/10.1002/jsfa.10138
4. de Souza PC, de Figueiredo FG, Honório AR et al (2019) Functional emulsion gels as pork back fat replacers in Bologna sausage. Food Struct 20:100105. https://doi.org/10.1016/j.foostr.2019.100105
5. de Carvalho FAL, Munekata PES, Pateiro M et al (2020) Effect of replacing backfat with vegetable oils during the shelf-life of cooked lamb sausages. LWT 122:109052. https://doi.org/10.1016/j.lwt.2020.109052
6. Öztürk-Kerimoğlu B, Kara A, Urgu-Öztürk M, Serdaroğlu M (2021) A new inverse olive oil emulsion plus carrot powder to replace animal fat in model meat batters. LWT 135:110044. https://doi.org/10.1016/j.lwt.2020.110044
7. Domínguez R, Lorenzo JM, Pateiro M et al (2024) Main animal fat replacers for the manufacture of healthy processed meat products. Crit Rev Food Sci Nutr 64(9):2513–2532. https://doi.org/10.1080/10408398.2022.2124397
8. Botella-Martínez C, Sayas-Barberá E, Pérez-Álvarez JÁ et al (2022) Chia and hemp oils-based gelled emulsions as replacers of pork backfat in burgers: effect on lipid profile, technological attributes and oxidation stability during frozen storage. Int J Food Sci Technol 58:3234-3243. https://doi.org/10.1111/ijfs.15907
9. Cerrón-Mercado F, Botella-Martínez CM, Salvá-Ruíz BK et al (2022) Effect of gelled emulsions elaborated with soybean oil, maca (*Lepidium meyenni*) four, and chincho (*Tagetes elliptica* S.) essential oil upon animal fat substitution in beef burgers. Foods 11:2198. https://doi.org/10.3390/foods11152198
10. Pintado T, Ruiz-Capillas C, Jiménez-Colmenero F et al (2015) Oil-in-water emulsion gels stabilized with chia (*Salvia hispanica* L.) and cold gelling agents: technological and infrared spectroscopic characterization. Food Chem 185:470–478. https://doi.org/10.1016/j.foodchem.2015.04.024
11. Domínguez R, Munekata PE, Pateiro M et al (2021) Immobilization of oils using hydrogels as strategy to replace animal fats and improve the healthiness of meat products. Curr Opin Food Sci 37:135–144. https://doi.org/10.1016/j.cofs.2020.10.005

12. Vargas-Ramella M, Munekata PES, Pateiro M et al (2020) Physicochemical composition and nutritional properties of deer burger enhanced with healthier oils. Foods 9:571. https://doi.org/10.3390/foods9050571
13. Carvalho Barros J, Munekata PES, de Carvalho FAL et al (2020) Use of tiger nut (*Cyperus esculentus* L.) oil emulsion as animal fat replacement in beef burgers. Foods 9:44. https://doi.org/10.3390/foods9010044
14. Ferreira I, Vasconcelos L, Leite A et al (2022) Use of olive and sunflower oil hydrogel emulsions as pork fat replacers in goat meat burgers: fat reduction and effects in lipidic quality. Biomolecules 12:1416. https://doi.org/10.3390/biom12101416
15. Botella-Martínez C, Gea-Quesada A, Sayas-Barberá E et al (2022) Improving the lipid profile of beef burgers added with chia oil (*Salvia hispanica* L.) or hemp oil (*Cannabis sativa* L.) gelled emulsions as partial animal fat replacers. LWT 161:113416. https://doi.org/10.1016/j.lwt.2022.113416
16. Commission Regulation (EC) No 148/2007 of 15 February (2007) Registering certain names in the Register of protected designation of origin and protected geographical indications. The OJEU 16/2/2007, L46/14
17. Commission Regulation (EC) No 676/2008 of 16 July (2008) Registering certain names in the Register of protected designations of origin and protected geographical indications. The OJEU 17/7/2008, L189/19.
18. Commission Implementing Regulation (EU) 2016/163 of 28 January (2016) Approving non-minor amendments to the specification for a name entered in the register of protected designations of origin and protected geographical indications. The OJEU 9/2/2016, L32/29
19. Botella-Martínez C, Fernández-López J, Ferreira I et al (2023) Alheiras with animal fat replacement: application of a gelled emulsion based on hemp oil (*Cannabis sativa* L.) and buckwheat. Eur Food Res Technol 249:2273–2285. https://doi.org/10.1007/s00217-023-04295-w

Check for updates

Chapter 3

Production of Low-Fat Bologna-Type Sausages Using Oleogel Rich in Oleic Acid

Suelem Lima da Silva, José Manuel Lorenzo Rodriguez, Alexandre José Cichoski and Paulo Cezar Bastianello Campagnol

Abstract

This chapter provides a detailed protocol for the production of low-fat Bologna-type sausages by replacing 50% of pork back fat with an oleogel composed of high oleic sunflower oil (HOSO), pork skin, and water. The oleogel, rich in oleic acid, was incorporated into the sausage formulation to enhance the lipid profile while maintaining technological and sensory quality. The protocol includes the step-by-step preparation of oleogel, sausage manufacturing procedures, cooking, packaging, and storage. This guide enables researchers and professionals to replicate a healthier sausage processing technique with reduced fat and cholesterol content.

Keywords Bologna sausage, Oleogel, Fat replacer, High oleic sunflower oil, Pork skin, Low-fat meat products

1 Introduction

Bologna-type sausages are widely consumed emulsified meat products, traditionally made with high levels of animal fat [1]. This contributes to an unfavorable fatty acid profile and high caloric content, factors linked to an increased risk of cardiovascular diseases. The development of healthier formulations by reducing animal fat has become a priority. However, reducing pork back fat poses challenges related to texture, flavor, and emulsion stability [2].

The use of oleogels—structured systems of vegetable oils—offers a promising solution for fat replacement. High oleic sunflower oil (HOSO) is particularly attractive due to its mild flavor, low cost, and health-promoting oleic acid content. When combined with pork skin, which acts as a natural gelling agent due to its high collagen content, it is possible to produce an oleogel with technological and nutritional functionality [3].

José Manuel Lorenzo Rodriguez (ed.), *Meat Product Quality Improvement*, Methods and Protocols in Food Science, https://doi.org/10.1007/978-1-0716-5460-6_3,

This chapter presents a standardized protocol for producing Bologna-type sausages with 50% of pork back fat replaced by an oleogel composed of HOSO, pork skin, and water.

2 Materials

2.1 Ingredients

- Lean pork meat (*Triceps brachii*)—65%
- Pork back fat—10%
- Oleogel (pork skin:water:HOSO in 1.5:1.5:1 ratio)—replaces 10% pork back fat (*see* **Note 1**)
- Ice—11.86%
- Sodium chloride—2.5%
- Sodium tripolyphosphate—0.3%
- Sodium nitrite—0.015%
- Sodium erythorbate—0.025%
- Coriander—0.20%
- Black pepper—0.10%

2.2 Equipment

- Cooking tank (80°C capacity)
- Bowl cutter
- Meat grinder with 3-mm plate
- Stuffing machine
- Ice bath
- Vacuum packaging machine
- Refrigeration chamber (4°C)

3 Methods

3.1 Oleogel Preparation

1. Cook pork skin at 80°C for 40 min.
2. Comminute the cooked pork skin in a blender until a uniform paste is obtained.
3. Mix pork skin paste, water, and HOSO in a ratio of 1.5:1.5:1 using a food processor (Fig. 1).
4. Store the oleogel overnight at 4°C to stabilize (*see* **Note 2**).

3.2 Sausage Manufacture

1. Grind lean pork, pork back fat, and oleogel separately using a meat grinder with a 3-mm plate.

Fig. 1 Visual aspect of oleogel produced with pork skin:water:HOSO in a 1.5:1.5:1 ratio

2. In a bowl cutter, emulsify lean pork with sodium chloride and sodium tripolyphosphate at high speed for 3 min to extract myofibrillar proteins.
3. Gradually add ice, sodium nitrite, sodium erythorbate, coriander, and black pepper. Mix at low speed for 2 min.
4. Add pork back fat and oleogel, and emulsify at high speed for two additional minutes. Ensure that the batter temperature remains below 12°C (*see* **Note 3**).
5. Stuff the emulsified batter into plastic casings (60 mm diameter, 400 g each) impermeable to water.

3.3 Cooking and Packaging

1. Cook the sausages in water in three steps: 60°C for 30 min, 70°C for 30 min, and 80°C until internal temperature reaches 72°C.
2. Cool immediately in an ice bath to 4°C.
3. Vacuum pack the sausages in oxygen- and moisture-barrier bags and store at 4°C (*see* **Note 4**).

4 Notes

1. The oleogel is prepared using a fixed ratio (1.5:1.5:1) of pork skin, water, and high oleic sunflower oil. This composition provides both gelling and emulsifying functions.
2. Refrigerated stabilization of the oleogel ensures appropriate consistency and thermal resistance during sausage processing.

3. Emulsion temperature control (<12°C) is critical to maintain emulsion stability and prevent protein denaturation.
4. Proper cooling and vacuum packaging prevent microbial growth and extend shelf life.

References

1. Feiner, G, 2006. Meat products handbook: Practical science and technology. Woodhead Publishing Limited, Cambridge, p. 648 ISBN978-1-84569-050-2
2. Anderson ET, Berry BW (2001) Effects of inner pea fiber on fat retention and cooking yield in high fat ground beef. Food Res Int 34(8):689–694. https://doi.org/10.1016/S0963-9969(01)00089-8
3. da Silva SL, Amaral JT, Ribeiro M et al. (2019) Fat replacement by oleogel rich in oleic acid and its impact on the technological, nutritional, oxidative, and sensory properties of Bologna-type sausages. Meat Sci 149:141–148. https://doi.org/10.1016/J.MEATSCI.2018.11.020

Chapter 4

Elaboration of Low-Sodium Dry-Cured Rabbit Leg

Douglas Pedro, José Manuel Lorenzo Rodriguez, Mirian Pateiro, Rubén Domínguez-Valencia, Bibiana Alves Dos Santos, Alexandre José Cichoski and Paulo Cezar Bastianello Campagnol

Abstract

This chapter describes a detailed protocol for the development of dry-cured rabbit legs with reduced sodium content through the partial replacement of sodium chloride (NaCl) with potassium chloride (KCl) and the addition of monosodium glutamate (MG) as a flavor enhancer. The protocol includes raw material selection, curing, drying, and storage steps. This guide enables researchers and industry professionals to reproduce the process and develop healthier alternatives to traditional dry-cured meat products.

Keywords Dry-cured meat, Rabbit leg, Sodium reduction, Potassium chloride, Monosodium glutamate, Functional meat products

1 Introduction

Rabbit meat is gaining attention due to its excellent nutritional profile, including high protein and essential amino acid content, low fat and cholesterol, and the presence of beneficial micronutrients, such as selenium and B vitamins [1]. Despite these attributes, the rabbit meat industry remains underdeveloped, primarily due to limited product variety and low consumer familiarity. Dry-cured rabbit leg represents an innovative approach to diversifying rabbit meat products inspired by traditional pork-based dry-cured hams.

Dry-cured meats are typically characterized by high sodium content, which plays a crucial role in flavor development, microbial safety, and texture [2]. However, health concerns related to high sodium intake have encouraged the meat industry to explore sodium reduction strategies [3]. Potassium chloride is a common sodium substitute, but its use above certain levels may lead to undesirable sensory attributes [4]. The use of flavor enhancers, such as monosodium glutamate (MG) can mitigate these effects and improve consumer acceptance [5].

José Manuel Lorenzo Rodriguez (ed.), *Meat Product Quality Improvement*, Methods and Protocols in Food Science, https://doi.org/10.1007/978-1-0716-5460-6_4,

This chapter outlines the methodology for elaborating low-sodium dry-cured rabbit legs with reduced sodium chloride content, partial substitution with potassium chloride, and inclusion of monosodium glutamate as a flavor enhancer. It provides detailed procedures and critical control points to ensure reproducibility and product quality.

2 Materials

Ingredients:

- Fresh rabbit legs (New Zealand White breed, ~200 g/leg, pH 5.4–5.7 within 2 h post mortem) (*see* **Note 1**)
- Sodium chloride (NaCl)—2.0% (w/w)
- Potassium chloride (KCl)—2.0% (w/w)
- Monosodium glutamate (MG)—0.3% (w/w)
- Sodium nitrite ($NaNO_2$)—100 ppm (0.01% per leg)
- Sodium nitrate ($NaNO_3$)—200 ppm (0.02% per leg)

Equipment:

- Climate-controlled chamber capable of operating at 2 ± 1°C with 85%–95% RH (for curing phase) and at 12 ± 1°C with 65%–75% RH (for drying phase)
- Vacuum packaging machine
- Weighing balance (precision ±0.1 g)
- Food-grade trays and gloves
- Thermometer and hygrometer
- Freezer (−18°C)

3 Methods

3.1 Raw Material Preparation

1. Select rabbit carcasses from animals weighing approximately 2.5–3.0 kg. Legs should be separated within 2 h post-mortem, with pH values between 5.4 and 5.7 (*see* **Note 1**).
2. Store the whole carcass at 2 ± 1°C for 12 h to complete rigor mortis.
3. Detach the hind legs using a sanitized knife.
4. Rinse each leg under running potable water for approximately 1 min to remove blood residues.
5. Hang legs vertically for 2 min to drain excess surface water.

6. Individually vacuum-pack the legs and freeze at −18°C for 30 days to standardize raw material and eliminate possible parasites (*see* **Note 2**).

3.2 Thawing and Salting

1. Thaw frozen rabbit legs at 2 ± 1°C and 80%–90% RH for 72 h in a refrigerated chamber.
2. Weigh each thawed leg. Based on individual weight, calculate the required quantities of each curing component:
   - Sodium chloride: 2.0% of leg weight
   - Potassium chloride: 2.0% of leg weight
   - Monosodium glutamate: 0.3% of leg weight
   - Sodium nitrite: 100 ppm (0.01% of leg weight)
   - Sodium nitrate: 200 ppm (0.02% of leg weight)
3. Manually rub the dry curing mix evenly onto all leg surfaces using gloved hands (*see* **Note 3**). Place each leg in individual food-grade trays to avoid cross-contamination.

3.3 Curing and Drying

1. Hang the salted legs on stainless steel hooks or nets in a curing chamber at 2 ± 1°C and 85%–95% RH for 7 days. Ensure sufficient spacing to allow uniform airflow.
2. After curing, transfer the legs to a drying chamber maintained at 12 ± 1°C and 65%–75% RH. Dry for 21 days to achieve moisture reduction and flavor development (Fig. 1).
3. Inspect legs daily for surface conditions. Remove any excess salt or moisture manually, if necessary (*see* **Note 4**).

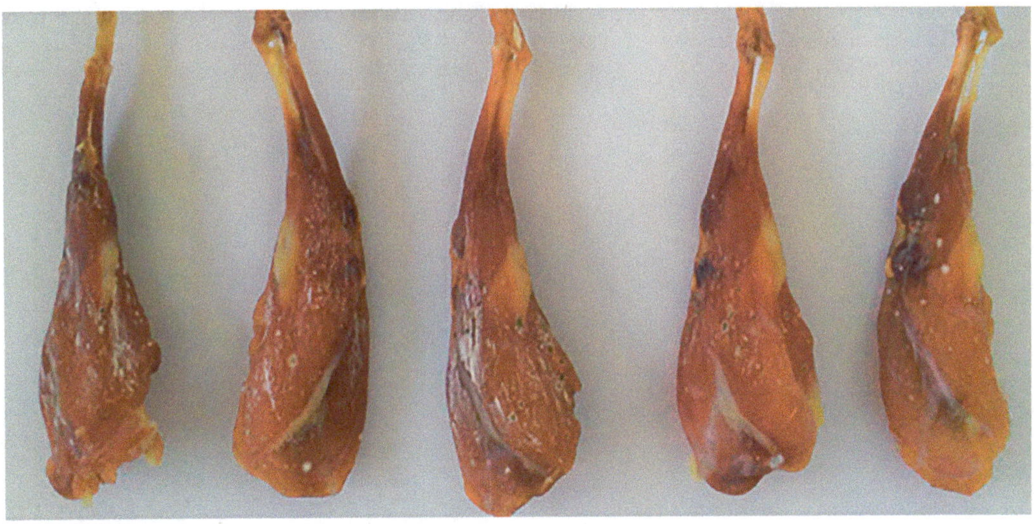

Fig. 1 Visual aspect of the dry-cured rabbit leg at the end of the manufacturing period

3.4 Packaging and Storage

1. At the end of the 28-day process (7 days curing + 21 days drying), vacuum-pack the legs in high-barrier plastic bags.
2. Store vacuum-packed products at 4°C for long-term storage (*see* **Note 5**).

4 Notes

1. Selecting rabbit legs with optimal post-mortem pH ensures proper curing and microbial safety.
2. Freezing improves batch uniformity and helps mitigate biological hazards.
3. Proper and uniform distribution of the dry mix is critical to prevent salt burns and ensure a balanced flavor.
4. Monitor temperature and humidity twice daily. Deviations may cause case hardening or spoilage.
5. Avoid prolonged storage above 4°C to prevent microbial growth and lipid oxidation.

References

1. Cullere M, Dalle Zotte A (2018) Rabbit meat production and consumption: state of knowledge and future perspectives. Meat Sci 143:137–146
2. Toldrá F (2010) Handbook of meat processing
3. Amine EK, Baba NH, Belhadj M, et al (2003) Diet, nutrition and the prevention of chronic diseases. In: World Health Organ Technical Report Series
4. Armenteros M, Aristoy MC, Barat JM, Toldrá F (2012) Biochemical and sensory changes in dry-cured ham salted with partial replacements of NaCl by other chloride salts. Meat Sci 90(2):361–367. https://doi.org/10.1016/j.meatsci.2011.07.023
5. dos Santos BA, Campagnol PCB, Morgano MAÔ, Pollonio MAR (2014) Monosodium glutamate, disodium inosinate, disodium guanylate, lysine and taurine improve the sensory quality of fermented cooked sausages with 50% and 75% replacement of NaCl with KCl. Meat Sci 96:509–513. https://doi.org/10.1016/J.MEATSCI.2013.08.024

Chapter 5

Elaboration of Dry-Cured Foal Cecina Using Kombu Seaweed as a Partial NaCl Replacer

Aurora Cittadini, Rubén Domínguez-Valencia, Rubén Agregán, Noemí Echegaray, Mirian Pateiro and José Manuel Lorenzo Rodriguez

Abstract

"Cecina" represents a typical Spanish dry-cured meat delicacy distinguished by its characteristic sensory attributes as intense color, firm texture, and distinctive flavor. However, the elevated sodium content typically associated with traditional dry-cured meat products constitutes a significant public health concern, thereby prompting both the meat industry and the scientific society to investigate effective sodium-reduction strategies. This chapter offers a detailed protocol for the production of foal "cecina" by replacing 50% NaCl with *Laminaria ochroleuca* (kombu) seaweed. In response to growing consumer demand for low salt meat products, this approach proposes the use of a mineral-rich brown seaweed powder as a natural and innovative salt replacer. This strategy enables substantial sodium reduction while enhancing the mineral profile of the final product without negatively affecting its physicochemical or sensory properties. Overall, this methodology demonstrates to be a promising technological strategy for developing a healthier traditional meat product while preserving its characteristic quality attributes. Thus, this book chapter details the raw material selection, formulation design, brine preparation, processing steps, and practical recommendations for the elaboration of this reformulated dry-cured "cecina".

Keywords Sodium reduction, *Laminaria ochroleuca* seaweed, Foal meat, Dry-cured meat product, Salt replacement

1 Introduction

"Cecina" is a traditional Spanish dry-cured meat product widely produced in northwestern Spain. The best-known variety, "Cecina de León," holds Protected Geographical Indication (PGI) status and is traditionally manufactured from beef hindquarters. However, beyond this specific PGI product, "cecina" can be also elaborated from other animal species, including foal, small ruminants (mainly goat), and game (such as venison). Regardless of the species used, this dry-cured meat product is characterized by unique organoleptic traits, as an intense flavor, firm texture, and distinctive aroma, and color [1]. Sodium chloride constitutes the cornerstone of "cecina"

José Manuel Lorenzo Rodriguez (ed.), *Meat Product Quality Improvement*, Methods and Protocols in Food Science,
https://doi.org/10.1007/978-1-0716-5460-6_5,

manufacture since it ensures microbial stability, contributs to texture and flavor development, as well as modulates key biochemical reactions responsible for the typical characteristics of dry-cured products [2–4]. Despite its technological importance, the sodium content of "cecina," as in most cured meat products, contributes to the dietary sodium intake. A moderate assumption of this mineral is essential for physiological functions, although it is recognized that an excessive ingestion (> 2 g sodium/day) is linked to increased blood pressure, which increments the risks of cardiovascular diseases, renal disorders, osteoporosis, stomach cancer, among others [6, 5]. In this context, the World Health Organization has established a global target of reducing population salt intake by 30%, recommending consumption below 2 g of sodium per day [6].

These public health objectives, together with increasing consumer demand for healthier food options, have prompted the exploration and development of innovative salt-reduction strategies within the meat sector, particularly in dry-cured varieties like "cecina" [7–9]. Among the different approaches investigated, total or partial substitution of NaCl with alternative ingredients (mineral salts) is one of the most employed. Nevertheless, though other chloride salts (as KCl, $CaCl_2$, and $MgCl_2$) and their mixtures have been widely investigated in other dry-cured products [2, 10–12], research specifically addressing salt reformulation in dry-cured "cecina" remains limited [13, 14]. In recent years, seaweeds, highly demanded for their potential beneficial effects, have emerged as encouraging natural and healthy ingredients for sodium reduction strategies [7, 15]. Indeed, these marine-derived components, representing excellent sources of minerals, such as potassium, calcium, magnesium (apart for sodium), as well as confers a salty taste due to their mineral richness. Among them, *Laminaria ochroleuca* (kombu), a brown macroalga widely consumed, stands out as a promising natural NaCl substitute, owing to its elevated dietary fiber and mineral contents as well as umami-enhancing properties [16, 17]. In parallel, foal meat represents an interesting raw material for the development of value-added dry-cured products. Actually, it is well-known for its favorable nutritional profile and environmental sustainability, aligning with current market trends toward healthier and environmentally conscious meat products [8, 18, 19].

Within this framework, this chapter aims to provide a comprehensive and reproducible protocol for the elaboration of dry-cured foal "cecina" using kombu seaweed as NaCl partial (50%) replacer [19]. This process ensures to obtain a "reduced in sodium" equine meat product with an improved mineral and nutritional profile, without negatively affecting most of the technological and sensory properties of the product. In this sense, integrating a valuable natural salt alternative in a nutritionally valuable raw material, this approach contributes to the development of a healthier and more sustainable traditional meat product. The chapter is intended to support researchers and industry professionals seeking practical solutions for sodium reduction without compromising product identity and quality.

2 Materials

According to the common production process, "cecina" requires only a limited number of essential elements: the selected meat cut, the salting system (brine or dry salt), and the appropriate infrastructures [20].

2.1 Raw Materials and Ingredients/Ingredients

This reformulated dry-cured foal *cecina* is manufactured from a whole-muscle cut subjected to brine salting. A whole foal knuckle corresponding to the *quadriceps femoris* muscle group (*vastus lateralis*, *vastus intermedius*, *vastus medialis*, and *rectus femoris* muscles), weighing 2.1 ± 0.5 kg (per piece), is used (*see* **Note 1**). This anatomical cut provides sufficient muscle mass and structural homogeneity to ensure controlled salt diffusion and gradual dehydration during the curing process (*see* **Note 2**).

For the brine solution the following ingredients are required:

- Common salt (NaCl), generally coarse sea salt
- Dehydrated kombu seaweed powder (*Laminaria ochroleuca*) obtained from a commercial supplier (*see* **Note 3**)
- Potable water

2.2 Material and Equipment

Specific processing materials and equipment are required for the elaboration of this product. The following materials are necessary for the brine preparation:

- Digital bench scale
- Graduated cylinder or volumetric flask
- Food-grade salting vats
- Stick blender or equivalent high-shear homogenizer device

While, for the "cecina" processing and curing steps, the main equipment includes:

- Knife
- Digital bench scale
- Identification tags (for batch traceability)
- Food-grade brush
- Salting room or refrigerated chamber (temperature-controlled)

- Post-salting chamber with controlled temperature and relative humidity
- Cooking-smoking chamber with temperature and relative humidity control
- Dry-curing chamber

3 Methods

The elaboration of dry-cured foal "cecina" using kombu seaweed as a partial NaCl replacer follows a controlled multistage process adapted from traditional "cecina" manufacture [20, 21] and adjusted to incorporate *Laminaria ochroleuca* powder as a mineral salt substitute [19]. The production steps are shown in Fig. 1 and described in detail below. Moreover, the processing conditions corresponding to each stage are summarized in Table 1.

3.1 Brine Preparation

Prepare a salt mixture composed of 50% sodium chloride (NaCl) and 50% Kombu seaweed powder (w/w). Formulate the brine so that the total dry salt mixture (NaCl + kombu powder) represents 30% (w/w)

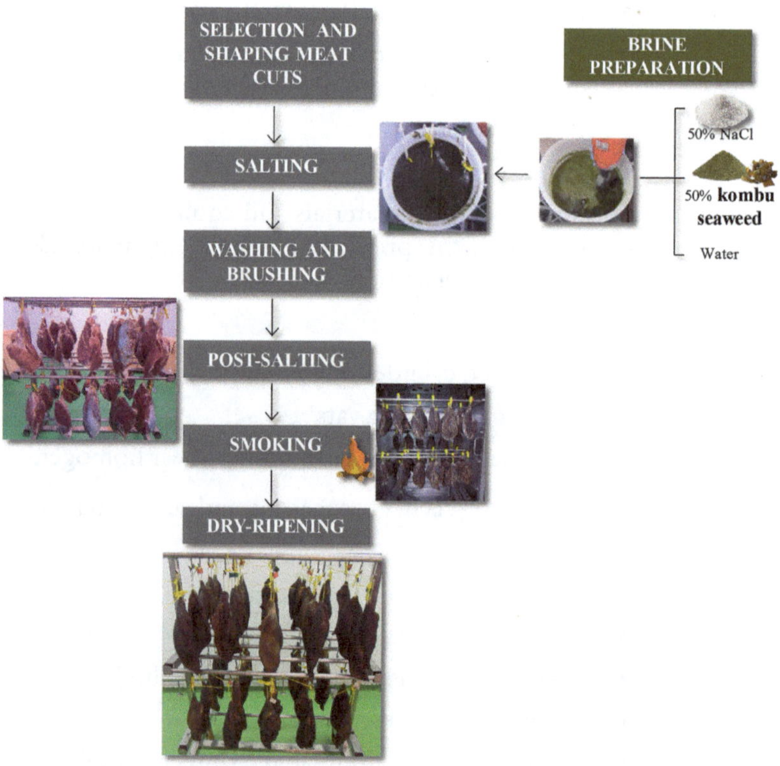

Fig. 1 Production process of dry-cured foal "cecina" using kombu seaweed as a partial NaCl replacer

Table 1
Processing conditions applied during dry-cured foal"cecina" production

Phase	Time	T (°C)	RH (%)
Salting	0.3 day/kg	2–5	85–90
Post-salting	60 days	2–5	85–90
Dry-smoking	2 h	25	–
Dry-ripening	15 days	8–10	75–80
Dry-ripening	45 days	12–14	70–75
Dry-ripening	45 days	12–14	65–70

T temperature, *RH* relative humidity

of the final brine, while potable water accounts for 70%. Gradually incorporate the dry salt mixture into potable water (2°C–5°C) under continuous agitation using a stick blender until complete dissolution and a uniform dispersion are achieved. Use the brine immediately after preparation or store it under refrigeration (2°C–5°C) for no longer than 24 h. Maintain brine temperature between 2°C and 5°C during salting to ensure microbiological safety and mineral stability.

3.2 Selection and Shaping of Meat Cuts

Select whole foal knuckles with uniform size and morphology. Trim each piece manually using a boning knife to remove excess external fat and large connective tissue structures. This operation improves external appearance and facilitates a homogeneous salt penetration and dehydration. Carry out this process in a suitable refrigerated cutting room. After shaping, when multiple pieces are processed simultaneously, label each knuckle with an identification code to ensure proper traceability and monitoring throughout the production process. Record the fresh weight of each piece in order to allow monitoring of weight loss throughout processing (*see* **Note 4**). Maintain the cuts under refrigeration until the salting stage.

3.3 Salting

Completely immerse each shaped cut in the prepared brine, ensuring full coverage. Salt each piece for 0.3 days/kg of fresh weight at 2°C–5°C with relative humidity (RH) conditions of 85%–90% (*see* **Note 5**). When processing multiple "cecinas," calculate salting time individually. At the beginning of the salting stage, gently massage the brine onto the meat surface to promote adhesion and initiate salt diffusion (*see* **Note 6**).

3.4 Washing and Brushing

After salting, wash the pieces thoroughly under running potable water (12°C–15°C) to remove excess surface salt and seaweed residues. Use soft food-grade brushes when necessary to eliminate adhered crystals. Allow excess surface water to drain before transferring the pieces to the post-salting chamber.

3.5 Post-Salting

Weigh the washed pieces and hang them individually with adequate spacing to allow proper air circulation. Transfer the meat cuts into a ventilated post-salting (settling) room, where remained for 60 days at 2°C–5°C and 85%–90% RH to avoid microbial growth (*see* **Note 7**). During this step, the brine, initially located near the surface of the cut, penetrates into the interior of the meat. Simultaneously, moisture migrates from the interior to the surface, contributing to gradual weight loss and reduction of water activity (*see* **Note 8**). Monitor weight at mid-stage and at the end of post-salting to assess curing progression.

3.6 Smoking

Transfer the pieces to a smoking chamber, where they are exposed to the smoke generated by the slow combustion of local oak wood, imparting a characteristic flavor and acting as a natural preservative. Apply dry smoking at approximately 25°C for 2 h (*see* **Note 9**).

3.7 Dry-Ripening

Transfer the smoked pieces to a controlled drying-ripening chamber with gentle and continuous air circulation (0.1–0.2 m/s). Apply a progressive temperature-humidity program: 15 days at 8°C–10°C and 75%–80% RH (for controlled moisture removal); 45 days at 12°C–14°C and 70%–75% RH (for intensification of enzymatic activity); and other 45 days at 12°C–14°C and 65%–70% RH (for final moisture reduction and flavor maturation) (*see* **Note 10**). Monitor the pieces periodically for weight loss, surface appearance, and texture firmness. The dry-maturation phase is fundamental for the development of the typical sensorial traits of this product (*see* **Note 11**). Consider the process complete when pieces reach water activity (a_w) values ≤0.90 (guarantying microbial stability) (*see* **Note 12**) and the characteristic firmness (resistance to thumb pressure without excessive hardness).

3.8 Physicochemical and Mineral Characterization

The physicochemical parameters and mineral composition of the dry-cured "cecina" elaborated with the kombu seaweed (K50) are presented in Table 2 and compared with those of a control product elaborated with 100% NaCl (CON) (Fig. 2). These values provide an overview of the main technological and nutritional implications associated with the partial sodium replacement by kombu seaweed. From a compositional perspective, the K50 formulation exhibits lower moisture and ash contents and a concomitant increase in protein concentration. On the other hand, fat and pH values remain unaffected by the salt reformulation. Besides, although a slight increase in water activity (a_w) is observed in the reformulated product, final values remain lower than 0.9 and in line with those commonly reported in "cecinas" (0.70–0.90), which ensuring the microbial stability of the product at room temperature [20].

Table 2
Physicoquemical parameters and mineral composition of dry-cured foal cecinas

Physicochemical parameters	CON	K50
Moisture (g/100 g)	37.65	34.40
Fat (dry matter) (g/100 g)	6.16	6.19
Protein (dry matter) (g/100 g)	73.78	77.29
Ash (dry matter) (g/100 g)	18.79	13.22
pH	5.93	5.96
a_w	0.75	0.79
Mineral composition		
Ca (mg/100 g)	10.34	23.31
K (mg/100 g)	844.34	1598.93
Mg (mg/100 g)	36.60	45.64
Na (mg/100 g)	2739.66	1423.03

Treatments: CON—100% NaCl; K50—50% NaCl and 50% Kombu seaweed. a_w water activity

Fig. 2 Appearance of foal dry-cured foal "cecina." CON—100% NaCl; K50— 50% NaCl and 50% Kombu seaweed

Furthermore, in particular, partial substitution of NaCl with *Laminaria ochroleuca* powder leads to a substantial reduction in sodium content, reaching a diminution of 48% compared with the CON formulation. In this sense, the K50 "cecina" can be labeled as "reduced in sodium" as containing at least 25% less sodium than a comparable product [22]. In addition, kombu incorporation not only diminishes the sodium concentration, but also enriches the product with potassium, magnesium, and calcium, thereby improving its overall mineral profile. This outcome aligns with current nutritional recommendations promoting reduced sodium intake and increased potassium consumption to support cardiovascular health

[23]. Overall, this alternative formulation favors the obtention of a final product with enhanced mineral and healthier profile, without altering most of the technological and sensorial properties [19].

4 Notes

1. Foal meat cut used for "cecina" manufacture should present uniform size, adequate muscular conformation, and absence of visible defects (hematomas, excessive connective tissue, and abnormal pH). Moreover, the intramuscular fat content influences dehydration rate, and as a consequence, the texture and flavor development [20]. Therefore, animal age, breed, and production conditions should be similar within batches to ensure reproducibility.
2. Although the knuckle is selected in this protocol, other whole-muscle cuts with similar morphology may be used. The most commonly used cuts include the round (in Spanish *tapa*), rump (*cadera*), silver side and eye of round (*contra con redondo*), as well as the central part of the silverside corresponding to the *gluteobiceps* muscle (*centro de contra*) [20].
3. Adequate knowledge of the seaweed's compositional and functional characteristics is required prior to its use, as proximate composition, mineral profile, moisture content (<10%), particle size, and microbiological quality (counts within acceptable limits). Alternative species (e.g., *Laminaria digitata* or *Saccharina japonica*) may also be used, although replacement levels may require adjustment depending on their mineral composition.
4. Monitor weight at regular intervals throughout the entire process in order to monitor the moisture loss progression and to detect in advance processing deviations [20].
5. The salting time and the RH may range from 0.3 to 0.6 days/kg of fresh weight, and from 80% to 90%, respectively. However, a proper management of this phase depends on temperature, cut size, and fat content [20].
6. The seaweed powder may impart a slight greenish coloration to the meat surface immediately after salting. This effect progressively diminishes during washing and ripening as chlorophyll derivatives degrade.
7. Although 60 days of post-salting is established in this protocol, this phase may last between 30 and 170 days depending on piece size and salt diffusion rate [20, 21].
8. Surface exudate may occasionally be observed probably related to the water-binding properties of seaweed polysaccharides

[24]. Regular inspection and removal of excess exudate are recommended in order to facilitate proper salt diffusion and prevent surface irregularities.

9. The smoking stage is optional and its duration may vary according to the desired intensity, generally extending from a few hours (per day) over a total period of 3–15 days [20, 21].
10. This step may last between 75 days to 22 months on the base of the meat piece size [20, 21].
11. A slow and controlled ripening process, adapted to the characteristics of each foal cut, is essential for achieving the optimal "cecina" eating quality. During this stage, proteolysis and lipolysis driven by endogenous muscle enzymes progressively shape the flavor and texture of the final product. These reactions should proceed at controlled rates through the application of low temperatures and reduced water activity (a_w), ensuring proper sensory development while maintaining product stability [20].
12. During ripening, weight loss results primarily from water evaporation. This moisture reduction combined with the salt absorbed during the salting stage promotes a decrease in water activity (a_w) to values ≤ 0.90, thereby ensuring microbial stability of the final product [20].

Acknowledgments

Aurora Cittadini acknowledges the support of a "Juan de la Cierva" postdoctoral fellowship (JDC2023 051424-I) funded by the Spanish "Ministero de Ciencia, Innovación y Universidades" and the "Agencia Estatal de Investigación" MCIU/AEI/10.13039/501100011033 and the FSE+. Noemí Echegaray and Rubén Agregán acknowledge GAIN for their postdoctoral grants (IN606C-2025/001 and IN606C-2025/002, respectively). Rubén Agregán, Noemi Echegaray, Mirian Pateiro, and José M. Lorenzo are members of the Competitive Reference Group "FunFood" (Functional Food), funded by GAIN (grant number IN607A2023/01).

References

1. Fernández D, Menéndez RA, Sanz JJ, García-Fernández M, del C (2015) Fatty acid profile of two cured meat products: dry-cured ham and cecina. Nutr Hosp 32:367–372. https://doi.org/10.3305/nh.2015.32.1.8911
2. Armenteros M, Toldrá F, Aristoy MC, Ventanas J, Estévez M (2012) Effect of the partial replacement of sodium chloride by other salts on the formation of volatile compounds during ripening of dry-cured ham. J Agric Food Chem 60:7607–7615. https://doi.org/10.1021/jf3013772
3. Inguglia ES, Zhang Z, Tiwari BK, Kerry JP, Burgess CM (2017) Salt reduction strategies in processed meat products—a review. Trends Food Sci Technol 59:70–78. https://doi.org/10.1016/j.tifs.2016.10.016
4. Lorenzo JM, Fonseca S, Gómez M, Domínguez R (2015) Influence of the salting time on

physico-chemical parameters, lipolysis and proteolysis of dry-cured foal "Cecina". LWT—Food Sci Technol 60:332–338. https://doi.org/10.1016/j.lwt.2014.07.023
5. World Health Organization (WHO) (2026) Sodium reduction. https://www.who.int/news-room/fact-sheets/detail/sodium-reduction. Accessed 10 Feb 2026
6. He FJ, Tan M, MacGregor GA (2019) Salt and health. In: Reducing salt in foods. Beeren C, Groves K, Titoria PM (eds) 2nd edn. Elsevier Ltd. ISBN 9780081008904
7. Aprilia GHS, Kim HS (2022) Development of strategies to manufacture low-salt meat products—a review. J Anim Sci Technol 64:218–234. https://doi.org/10.5187/jast.2022.e16
8. Teixeira A, Rodrigues S (2021) Consumer perceptions towards healthier meat products. Curr Opin Food Sci 38:147–154. https://doi.org/10.1016/j.cofs.2020.12.004
9. Huang R, Fang Y, Lu W, Zhong Y, Deng Y, Zhang M, Zhao F (2025) Rethinking salt in dry-cured meats: innovations for a healthier and flavorful future. Food Rev Int 41:2437–2460. https://doi.org/10.1080/87559129.2025.2472237
10. Domínguez R, Munekata PE, Cittadini A, Lorenzo JM (2016) Effect of the partial NaCl substitution by other chloride salts on the volatile profile during the ripening of dry-cured lacón. Grasas y Aceites 67:e128. https://doi.org/10.3989/gya.0505152
11. Lorenzo JM, Cittadini A, Bermúdez R, Munekata PE, Domínguez R (2015) Influence of partial replacement of NaCl with KCl, CaCl2 and MgCl2 on proteolysis, lipolysis and sensory properties during the manufacture of dry-cured lacón. Food Control 55:90–96. https://doi.org/10.1016/J.FOODCONT.2015.02.035
12. Pedro D, Saldaña E, Lorenzo JM, Pateiro M, Dominguez R, Dos Santos BA, Cichoski AJ, Campagnol PCB (2021) Low-sodium dry-cured rabbit leg: a novel meat product with healthier properties. Meat Sci 173:108372. https://doi.org/10.1016/j.meatsci.2020.108372
13. Cittadini A, Domínguez R, Gómez B, Pateiro M, Pérez-Santaescolástica C, López-Fernández O, Sarriés MV, Lorenzo JM (2020) Effect of NaCl replacement by other chloride salts on physicochemical parameters, proteolysis and lipolysis of dry-cured foal "Cecina". J Food Sci Technol 57:1628–1635. https://doi.org/10.1007/s13197-019-04195-6
14. Vargas-Ramella M, Lorenzo JM, Domínguez R, Pateiro M, Munekata PES, Campagnol PCB, Franco D (2021) Effect of NaCl Partial replacement by chloride salts on physicochemical characteristics, volatile compounds and sensorial properties of dry-cured deer Cecina. Foods 10:669. https://doi.org/10.3390/foods10030669
15. Circuncisão AR, Catarino MD, Cardoso SM, Silva AMS (2018) Minerals from macroalgae origin: health benefits and risks for consumers. Mar Drugs 16:400. https://doi.org/10.3390/md16110400
16. Fernández-Segovia I, Lerma-García MJ, Fuentes A, Barat JM (2018) Characterization of Spanish powdered seaweeds: composition, antioxidant capacity and technological properties. Food Res Int 111:212–219. https://doi.org/10.1016/j.foodres.2018.05.037
17. Sharma U, Jadaun S, Khapudaang R, Siddiqui S (2024) Seaweed—a sustainable food source in the food industry. In: Thakur M (ed) Sustainable food systems, vol II. World sustainability series. Springer, Cham, pp 187–203
18. Belaunzaran X, Bessa RJB, Lavín P, Mantecón AR, Kramer JKG, Aldai N (2015) Horsemeat for human consumption—current research and future opportunities. Meat Sci 108:74–81. https://doi.org/10.1016/j.meatsci.2015.05.006
19. Cittadini A, Domínguez R, Sarriés MV, Pateiro M, Lorenzo JM (2023) Study of Pansalt® or *Laminaria ochroleuca* seaweed powder as potential NaCl replacers in dry-cured foal "cecina". Meat Sci 204:109253. https://doi.org/10.1016/j.meatsci.2023.109253
20. Ordéñez Gutiérrez N, Caro I, Mateo J (2022) Dry-cured Cecina. In: Lorenzo JM, Domínguez R, Pateiro M, Munekata PES (eds) Production of traditional mediterranean meat products. Springer, New York, USA. ISBN 9781071621035
21. Andrés S, Pazos A, Perlo F, Rosmini M, Aleu G, Campagnol P, dos Santos B, Trindade M, Contreras Castillo C, Sepúlveda N et al (2020) Catálogo de productos cárnicos iberoamericanos. CYTED, Madrid, Spain. ISBN 9789895443512
22. European Commission Regulation (EC) No 1924/2006 of the European Parliament and of the Council of 20 December 2006 on Nutrition and Health Claims Made on Foods (2006) Off J Eur Union L404:9–25
23. World Health Organization (WHO) Salt reduction. https://www.who.int/news-room/fact-sheets/detail/salt-reduction. Accessed 23 Feb 2023
24. Cosenza VA, Navarro DA, Ponce NMA, Stortz CA (2017) Seaweed polysaccharides: structure and applications. In: Industrial applications of renewable biomass products past, present and future. Nair Goyanes S, D'Accorso NB (eds) 1st edn. Springer Cham. ISBN 9783319612874

Check for updates

Chapter 6

Edible Mushrooms in the Elaboration of Reduced Salt and Fat Emulsions

Magdalena I. Ceron Guevara, Rubén Agregán, Esmeralda Rangel-Vargas and Eva M. Santos

Abstract

The reformulation of meat products through the incorporation of plant-based ingredients has gained particular relevance in the past decade. This is especially due to certain negative aspects attributed to meat products, such as their association with certain types of cancer and sustainability issues, particularly those related to red meat. In this regard, the edible mushroom can be a feasible ingredient to partially replace salt and fat in emulsions, such as frankfurters. In this chapter, the process to successfully incorporate the ingredient in emulsioned meat products is described.

Keywords Emulsions, Plant based foods, Salt and fat replacement

1 Introduction

Although meat and meat products are considered an excellent protein, iron, and vitamins source, two main concerns have reached consumer perception in the last decade. On one hand, negative health issues have been related to the consumption of meat products because of the presence of saturated fats, cholesterol levels, and high salt contents, which can be associated with certain illnesses, such as metabolic syndrome, obesity, and cardiovascular events [1, 3]. On the other hand, sustainability claims concerning the ecological footprint attributed to growth cattle are arising [4]. In this context, there have been an explosion of new alternatives in the food chain, from the replacement of certain ingredients (hybrid meat products) to the development of meat analogues, in the corriente of the plant based foods [5].

One of the new ingredients in the spotlight are the edible mushrooms. With more than 14,000 mushrooms known species and 3000 edible species, the mushrooms become a diverse and significant source to consider in the meat products formulation [6]. Edible

José Manuel Lorenzo Rodriguez (ed.), *Meat Product Quality Improvement*, Methods and Protocols in Food Science, https://doi.org/10.1007/978-1-0716-5460-6_6,

mushrooms are characterized by high moisture content, with an interesting presence of protein and fiber, especially when they are transformed into powder, which make them an excellent candidate to improve the nutritional profile of meat products [7]. However, the incorporation of nonmeat ingredients present sensorial challenges since color, texture, or flavor are affected [8]. The edible mushroom species *Agaricus bisporus* and *Pleurotus ostreatus* were tested as partial fat and salt replacers in several meat products, such patties, frankfurters, and pate [8–10]. In this chapter, the elaboration process of incorporating mushroom flour in emulsioned frankfurters is described.

2 Materials

In this case, the edible mushrooms *Agaricus bisporus* (Ab.) and *Pleurotus ostreatus* (Po) are used to elaborate the emulsions. The mushrooms can be purchased in local markets but it is important to assure the quality of the mushrooms in their appearance: light color, good size, and maturity level, as well as the absence of physical damage. Figure 1 shows the process of mushroom flour obtention. To elaborate the frankfurters the following ingredients are also required: frozen or refrigerated post rigor lean pork meat, pork backfat, phosphate blend for meat products, NaCl, sodium caseinate, water, potato starch, sodium nitrite, sodium ascorbate, and collagen casings of 20 mm.

3 Methods

3.1 Obtention of Mushroom Flour

The preparation of mushroom flour is relatively simple as shown in Fig. 1.

1. Wash the mushrooms with tap water to eliminate the dust, soil, and impurities from the material.
2. Drain the excess of water and cut in slices around 5 mm thickness.
3. In the case of *Agaricus bisporus* mushrooms, a scalding step in a 2% acetic acid solution at 80°C for 10 min with a posterior cooling in cold water at 4°C is necessary to avoid enzymatic reactions and browning. Drain the excess of water again (*see* **Note 1**).
4. Air drying of the mushroom slices in an electric oven with air circulation at 45°C ± 5°C for 24–48 h to reduce the water content of the product (Fig. 1) (*see* **Note 2**).

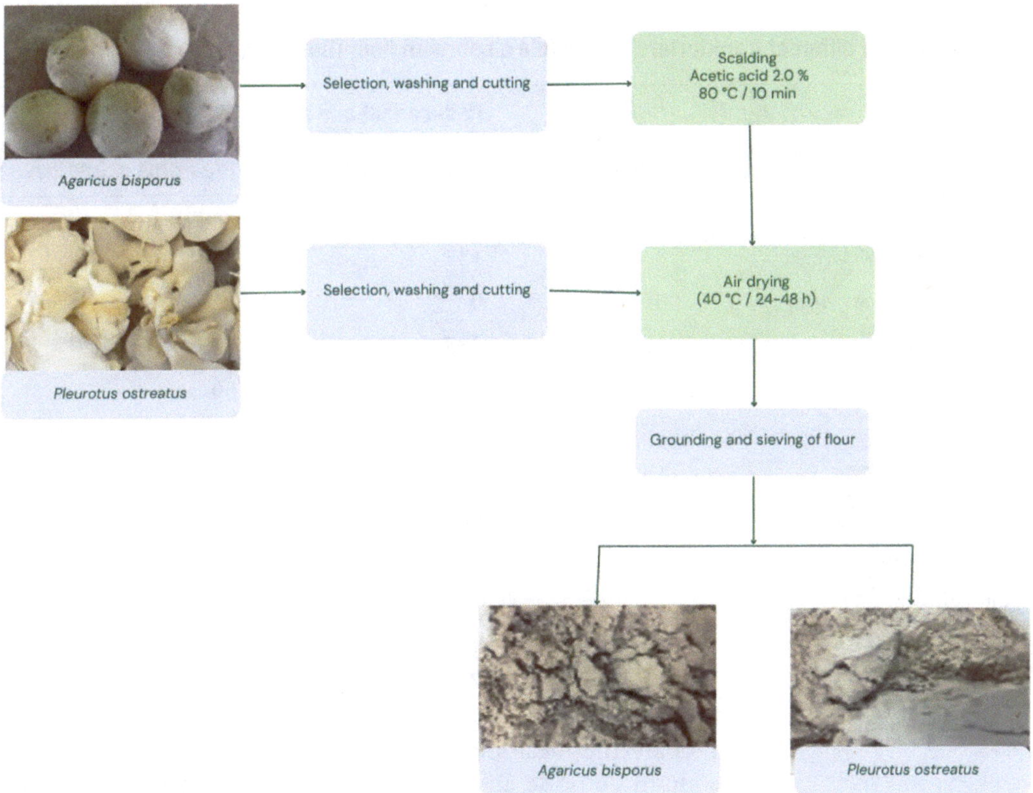

Fig. 1 Process of mushroom flour obtention

5. When the slices look dry and tough, they are ground to a particle size around 0.5 mm (Fig. 1) (*see* **Note 3**).
6. The flours are then stored in UV light protected airtight storage containers at room temperature until their use (*see* **Note 4**).

3.2 Elaboration Process of Frankfurters

Comparing with an standard frankfurter formulation where 25% of pork backfat and 1.5% of salt, in addition to 2% of sodium caseinate, 0.5% of phosphates blend, 1% potato starch, 0.03% sodium nitrite, and 0.05% sodium ascorbate are used, two formulations are proposed for the partial replacement of 30% of animal fat and 50% of the salt formulation by incorporating 2.5% of mushroom Ab or Po flour. Incorporations of higher percentages of mushroom flour result in darker color sausages. In Table 1, the formulation of the reduced fat and sodium frankfurter is presented to elaborate 5 kg (Table 1).

Figure 2 shows the process of elaboration of frankfurters. The process of elaboration of frankfurters is as follows.

1. Weight every ingredient according to the final weight of formulation.

Table 1
Formulation (g) of different frankfurters with edible mushroom flour (Batch 5 kg)

	Standard frankfurter	Reduced fat and salt frankfurters	
		Agaricus bisporus	*Pleurotus ostreatus*
Lean meat	2500	2500	2500
Flour *Agaricus bisporus*	0	125	0
Flour *Pleurotus ostreatus*	0	0	125
Pork backfat	1250	875	875
Sodium caseinate	100	50	50
Sodium chloride	75	37.5	37.5
Water	1000	1346	1346
Potato starch	50	50	50
di-tri phosphates	25	12.5	12.5
Sodium nitrite	1.5	1.5	1.5
Sodium ascorbate	2.5	2.5	2.5

2. Prepare the meat and the fat in cubes of 2.5 cm and add the sodium chloride and sodium nitrite to the meat with a rest period of 2 h (*see* **Note 5**).
3. Add potato starch and 50% of total sodium caseinate (dissolved in water) to meat. Mince the meat and fat using plates of 8 and 6 mm, respectively, in a refrigerated mincer machine (*see* **Note 6**).
4. In a chilled cutter hydrate the mushroom flour with water to get puree and add the meat, phosphates, and the rest of the water and process at regular speed. After obtaining the paste, add the fat and the rest of the ingredients and work the mixture at double speed to form the emulsion (*see* **Note 7**).
5. Finally, add the colorant and frankfurter flavor (if desired) and the sodium ascorbate dissolved in a little water.
6. Meat batter is stuffed into 20 mm collagen casings with an automatic equipment forming 10 cm frankfurters (*see* **Note 8**).
7. The raw sausages can be cooked by immersion in a hot water at 80°C for 30 min (*see* **Note 9**).
8. Then, the cooked sausages are cooled in an ice water-bath to reach 4°C. At this point, the product can be left overnight at refrigeration temperatures.

Fig. 2 Process of elaboration of frankfurters

9. Finally the frankfurters are vacuum packed in polyethylene plastic bags and pasteurized at 90°C for 30 min and stored at 4°C with a shelf life of 3 months [2, 5, 11].

4 Notes

1. If scalding is not applied to *Agaricus bisporus*, the subsequent air drying process will negatively affect the flour color, increasing the darkness of the powder and giving an undesirable emulsion color, affecting its acceptance.

2. Higher temperatures until 60°C can be used to dry the mushrooms, but higher temperatures accelerate the browning reactions affecting the color of the ingredient, and therefore, the frankfurters appearance.
3. Avoid overheating during the grounding process to demerit the quality of the flours and to avoid burnt flavors.
4. Limiting the UV light and oxygen in the package delays oxidation reactions extending the shelf-life of the flours.
5. If necessary the resting period of 2 h can be avoided although some coloration problems could be noticed.
6. The caseinate has to be dissolved in hot water (75°C).
7. The emulsion is the mixture of two or more immiscible liquids, where one is dispersed in the other in the form of small droplets. In the frankfurter, the small droplets of fat are dispersed in the aqueous phase formed by meat proteins, water, and additives, while myosin is acting as an emulsifier. The emulsion is ready, when it presents an homogeneous and uniform appearance texture and consistency.
8. Instead of collagen casings, cellulose casings can be used, but in this case, after the cooking and cooling processes, the sausages have to be peeled off.
9. The raw sausages can be also cooked in an electric oven 20 min at 50°C (without humidity), 20 min at 78°C with 100% of humidity to reach 72°C core temperature and finalize with a cold water shower.

Acknowledgments

Authors (E. R.-V., R. A., E.M.S.) are members of the AlProSos network (Sustainable Regional Protein-rich Foods), funded by CYTED (ref. 125RT0165). Rubén Agregán acknowledges GAIN for their postdoctoral grants (IN606C-2025/002). Rubén Agregán is member of the Competitive Reference Group "FunFood" (Functional Food), funded by GAIN (grant number IN607A2023/01).

References

1. Hathwar SC, Rai AK, Modi VK, Narayan B (2012) Characteristics and consumer acceptance of healthier meat and meat product formulations—a review. J Food Sci Technol 49:653–664
2. Ekmekcioglu C, Wallner P, Kundi M, Weisz U, Haas W, Hutter HP (2018) Red meat, diseases, and healthy alternatives: a critical review. Crit Rev Food Sci Nutr 58(2):247–261
3. Campos-González N, Gómez-Salazar JA, Cerón-García A, Lorenzo JM, Santos EM, Campagnol PC, Sosa-Morales ME (2024) Approaches to assess the functionality of meat products made from the incorporation of

vegetable by-products: a review. Food Rev Int 40(6):1678–1702
4. Parlasca MC, Qaim M (2022) Meat consumption and sustainability. Ann Rev Resour Econ 14(1):17–41
5. Grasso S (2024) Opportunities and challenges of hybrid meat products: a viewpoint article. Int J Food Sci Technol 59(11):8693–8696
6. Perez-Montes A, Rangel-Vargas E, Lorenzo JM, Romero L, Santos EM (2021) Edible mushrooms as a novel trend in the development of healthier meat products. Curr Opin Food Sci 37:118–124
7. Rangel-Vargas E, Rodriguez JA, Domínguez R, Lorenzo JM, Sosa ME, Andrés SC, Santos EM (2021) Edible mushrooms as a natural source of food ingredient/additive replacer. Foods 10(11):2687
8. Das AK, Nanda PK, Dandapat P, Bandyopadhyay S, Gullón P, Sivaraman GK, Lorenzo JM (2021) Edible mushrooms as functional ingredients for development of healthier and more sustainable muscle foods: a flexitarian approach. Molecules 26(9):2463
9. Cerón-Guevara MI, Rangel-Vargas E, Lorenzo JM, Bermúdez R, Pateiro M, Rodriguez JA, Santos EM (2020) Effect of the addition of edible mushroom flours (*Agaricus bisporus* and *Pleurotus ostreatus*) on physicochemical and sensory properties of cold-stored beef patties. J Food Process Preserv44(3):e14351
10. Cerón-Guevara MI, Rangel-Vargas E, Lorenzo JM, Bermúdez R, Pateiro M, Rodríguez JA, Santos EM (2020) Reduction of salt and fat in frankfurter sausages by addition of *Agaricus bisporus* and *Pleurotus ostreatus* flour. Foods 9(6):760
11. Cerón-Guevara MI, Santos EM, Lorenzo JM, Pateiro M, Bermúdez-Piedra R, Rodríguez JA, Rangel-Vargas E (2021) Partial replacement of fat and salt in liver pâté by addition of *Agaricus bisporus* and *Pleurotus ostreatus* flour. Int J Food Sci Technol 56(12):6171–6181

Check for updates

Chapter 7

Low Sodium Chicken Nuggets

Julliane Carvalho Barros and Marco Antonio Trindade

Abstract

Chicken nuggets are a staple in supermarkets and fast foods restaurants, which makes it a convenient food, that is, consumed by the entire population, regardless of income and age group. However, the amount of salt (NaCl) added in traditional formulations of chicken nuggets, until 2.5 g/100 g NaCl, increasing the sodium content in chicken nuggets. Furthermore, chicken nuggets with a low amount of sodium chloride may result in a decrease in the concentration of the extracted protein, involved in the formation of the gel/emulsion matrix. The objective of this chapter is to provide a detailed description of the steps involved in processing low-sodium chicken nuggets by replacing NaCl with $CaCl_2$ with the same ionic strength.

Keywords Sodium content, Calcium chloride, Ionic strength, Breaded meat product reformulation, Healthiness

1 Introduction

These days, chicken nuggets are a staple in supermarkets and fast foods restaurants. Chicken nuggets are considered breaded meat products, produced from ground chicken meat, with the addition of other ingredients, such as chicken skin, starch, and soy proteins, used to reduce their cost. In addition to its valuable nutrient content like other meat products, this product is available in a partially pre-fried form, which makes it a convenient food, that is, consumed by the entire population, regardless of income and age group [1].

Regarding the amount of salt (NaCl) added in traditional formulations of chicken nuggets, 2–2.5 g/100 g NaCl have been used, increasing the sodium content in this meat product [2, 3]. However, according to [4], high blood pressure is influenced by excessive sodium intake in foods. The World Health Organization [5] recommends that NaCl consumption be less than 5 g/day, which is equivalent to a maximum daily sodium intake (Na^+) of 2 g/day. Several studies have aimed to reduce the sodium content in meat products, such as sausages, chicken nuggets, and mortadella. This reduction in NaCl can occur through the simple reduction of NaCl or through

José Manuel Lorenzo Rodriguez (ed.), *Meat Product Quality Improvement*, Methods and Protocols in Food Science,
https://doi.org/10.1007/978-1-0716-5460-6_7,

partial or total replacement with other chlorinated salts, such as calcium ($CaCl_2$), potassium (KCl), and magnesium ($MgCl_2$) [6–10].

However, reducing NaCl in meat batter requires other ionic compounds to restore meat functionality, such as water retention (WHC), protein, and fat binding [11]. Because NaCl plays a significant role in extracting myofibrillar proteins from meat, it binds to water and fat, stabilizing the meat batter. Chicken nuggets with a low amount of sodium chloride may result in a decrease in the concentration of the extracted protein, involved in the formation of the gel/emulsion matrix [3].

Furthermore, reformulating meat products with reduced sodium content is not a simple task, as salt reduction not only affects the salty taste, but the intensity of the product in relation to the characteristic flavor [12]. Yogesh et al. [2], studying different concentrations of NaCl (2.5%–0%) and fat (5% and 0%) in chicken nuggets, found that salinity decreased with the reduction of NaCl, negatively affecting the palatability of chicken nuggets.

Verma et al. [13] evaluated the quality of low-fat chicken nuggets added with a mixture of salt substitutes (0.2% KCl, 1% sucrose, 0.03% citric acid, and 0.03% tartaric acid) and pea shell flour, verified that the mixture of salts decreased product yield and emulsion stability in low-fat chicken nuggets. In the study carried out by Barros et al. [14], replacing 75% of the added NaCl with $CaCl_2$, with the same ionic strength and equivalent to 1.5% NaCl in chicken nuggets, caused a reduction of up to 43% in sodium and did not affect cooking performance.

The objective of this chapter is to provide a detailed description of the steps involved in processing low-sodium chicken nuggets by replacing NaCl with $CaCl_2$ with the same ionic strength.

2 Material

The ingredients used to reformulate chicken nuggets with a 75% reduction in sodium chloride content and replaced with $CaCl_2$ with ionic strength equivalent to 1.5% NaCl are (*see* **Note 1**):

- 77.7 g/100 g chicken breast fillet,
- 20 g/100 g chicken skin,
- 0.375 g/100 g NaCl, sodium content reduced by 75% from a standard formulation of 1.5% NaCl,
- 0.712 g/ 100 g $CaCl_2$ (food grade) depending on the ionic strength equivalent to 1.5% NaCl,
- 0.05 g/100 g sodium erythorbate,
- 0.25 g/100 g sodium tripolyphosphate,
- 0.30 g/100 g onion powder,

- 0.15 g/100 g garlic powder,
- 0.05 g/100 g white pepper.

The breading flours are:

- Pre-dust (type PDA 66-TC; Kerry Company),
- Batter coating (adhesion type BMA91-TC; Kerry Company),
- Breading (dressing type SA F05; Kerry Company).

3 Methods

To reduce the sodium content of chicken nuggets by replacing sodium chloride with calcium chloride, without affecting their technological and sensorial properties, the substitution must be made depending on the ionic strength (IS) equivalent to the amount of NaCl to be used. The ISs of the different salt concentrations are calculated according to Segel [15]:

$$\mathrm{IS} = \frac{1}{2} \sum_{i-1}^{n} C_i Z_i^2 \tag{1}$$

Where:

- IS = ionic strength,
- C_i = molarity of the ion (M, mol/L),
- Z_i = the net charge of the ion,
- M = n1/V(L) unit: mol/L,
- n1 = m (grams of compound)/molar mass (g/mol),
- V = 1 L = 1 kg,
- NaCl Molar mass = 58.44 g/mol,
- $CaCl_2$ Molar mass = 110.98 g/mol,

The calculation of the ionic strength of 1.5% NaCl in a 1000 g meat batter containing 1.5% NaCl are:

$$1.5\text{g NaCl} - 100\text{g batter}$$

$$15\text{g NaCl} - 1000\text{g batter}$$

The number of moles of NaCl = (amount of NaCl in 1000 g/ NaCl molar mass)

$$= 15/58.44 = 0.256 \text{ mols}$$

$$M = 0.256/1 = 0.256 \text{ mol/L}$$

Using Eq. (1), the ionic strength (IS) for NaCl 1.5% is:

$$½[(0.256 \times 12) + (0.256 \times 12)] = 0.256$$

To calculate the ionic strength of $CaCl_2$ as a function of the 75% reduction in NaCl, we have:

$$FI(CaCl_2) = 0.256 \times 0.75 = 0.192$$

Therefore,

$$FI(CaCl_2) = 3 \times C(CaCl_2)$$

$$0.192 = 3 \times (m/110.98)$$

$$m = 7.10 g/kg = 0.710 g/100 g\ CaCl_2$$

As calculated, to reduce the amount of NaCl by 75% and replace it with $CaCl_2$ with ionic strength equivalent to 1.5% NaCl (IS = 0.256), a smaller amount of $CaCl_2$ is necessary, as can be seen in Table 1.

3.1 Processing of Chicken Nuggets

To process chicken nuggets (Fig. 1), the raw materials must be selected and the ingredients weighed, as previously mentioned. The skinless chicken breast must be ground in a meat grinder on an 8 mm disc and the chicken skin on a 4 mm disc. For better extraction of myofibrillar proteins, meat raw materials must be homogenized with sodium and calcium chloride salts. Subsequently, the other ingredients, sodium erythorbate, sodium tripolyphosphate, onion powder, garlic powder, and white pepper must be added to the meat batter until complete homogenization. After mixing all the ingredients, the meat batter is shaped into chicken nuggets and frozen for better adhesion to the coating system.

Table 1

$CaCl_2$ concentrations in chicken nuggets with ionic strength equivalent to 1.5% NaCl

	NaCl		$CaCl_2$	
Treatments	**g/100 g**	**IS[†]**	**g/100 g**	**IS[a]**
25%NaCl	0.375	0.064	0.710	0.192

[a] *IS* ionic strength

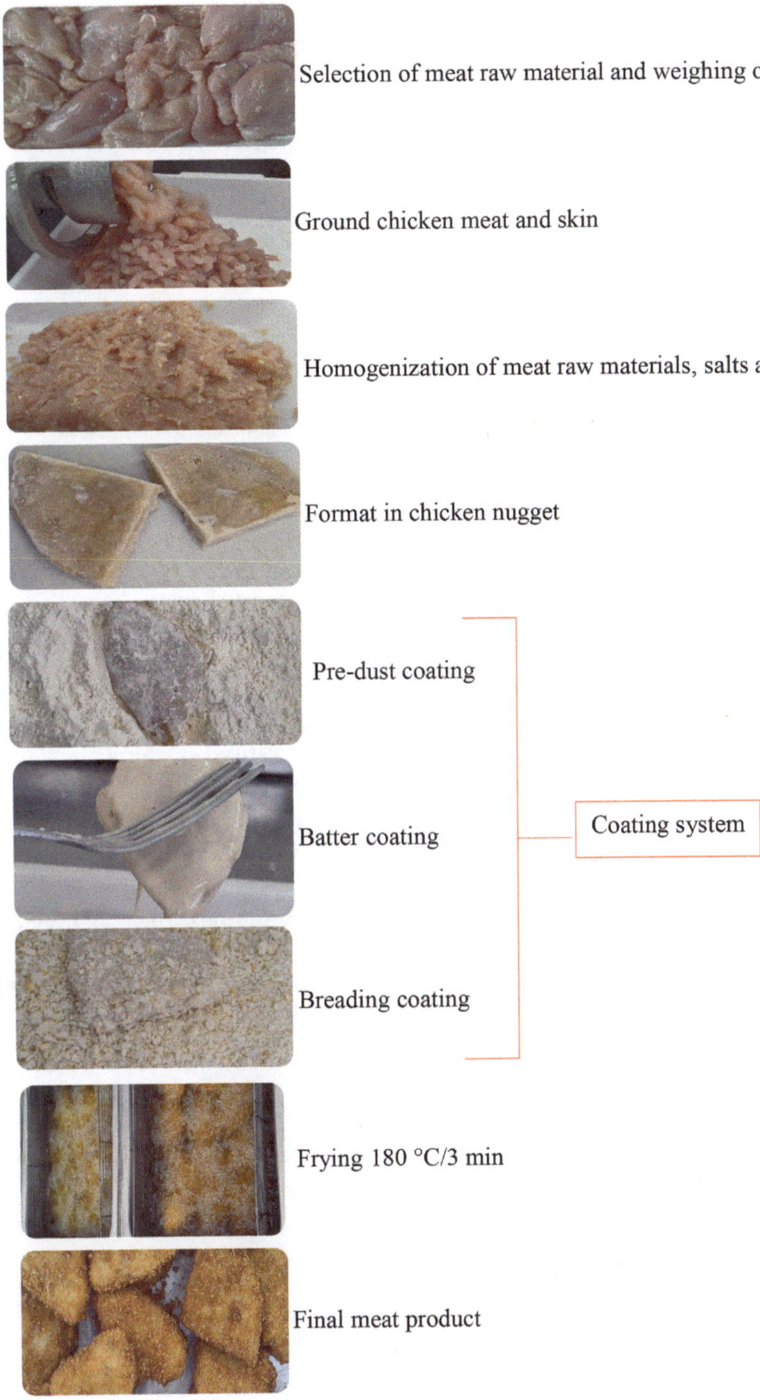

Fig. 1 Chicken nuggets processing flowchart

The stage following freezing is the breading of the chicken nuggets, which occurs according to the following steps (Fig. 2) (A) Pre-dust: the already shaped and frozen chicken nuggets pass

Fig. 2 Stages of breading chicken nuggets. (**A**) Pre-dust, (**B**) Batter coating, and (**C**) Breading

through a thin layer of flour, which is intended to adhere to the next breading stage, batter; (B) Batter coating: the shaped meat mass is immersed in a solution prepared in a 1:2 ratio (powder:water), which aims to improve the adhesion of the breading flour, and finally, we have breading; (C) Breading: is a flour (bread crumb) with larger particle size, which has the purpose of providing the crispness of the chicken nuggets [16].

During the breading of the chicken nuggets, the pick up yield is calculated, that is, how much coating was enough to coat the entire meat batter. The chicken nuggets were weighed before and after the three-step coating system (pre-dust, batter, and breading), and the pick-up was calculated by [17]:

$$\text{Pick} - \text{up} = \left(\frac{\text{Nuggets with Coating} - \text{Nuggets without coating}}{\text{Nuggets with Coating}}\right) \times 100$$

For the formulation of chicken nuggets with reduced sodium content and replaced by $CaCl_2$ with ionic strength equivalent to 1.5% NaCl, there is a pick up of around 32%, which was very close to the 30%, suggested by the coating system manufacturer.

After the coating steps, the chicken nuggets are fried, for example, in an electric fryer with a capacity of 2.5 L. At this stage, the chicken nuggets are fried in vegetable fat at 180°C until the internal temperature of the meat product is at least 72°C to ensure safety, and a skewer-type thermometer is used to check the internal temperature (*see* **Note 2**). Finally, the chicken nuggets are packaged in polyethylene plastic and stored under freezing at −18°C. Under these conditions, chicken nuggets have a cooking yield of around 92%, which is an important parameter for the food industry.

4 Notes

1. This formulation is capable of reducing the sodium content (~40%), without affecting the technological properties of the chicken nuggets.
2. It does not affect the cooking yield, as it maintains the same ionic strengths as NaCl when replaced NaCl by $CaCl_2$.

Acknowledgments

The authors gratefully acknowledge the Coordenação de Aperfeiçoamento de Pessoal de Nível Superior (CAPES)—Finance Code 001 (CAPES, Brazil) for the doctoral scholarship granted to the first author (JC Barros). Also to the Fundação de Amparo à Pesquisa do Estado de São Paulo [São Paulo Research Foundation (FAPESP, Brazil)] for supporting the research (Process 2015/12429-7).

References

1. Barbut S (2012) Convenience breaded poultry meat products—new developments. Trends in Food Sci Technol 26(1):14–20. https://doi.org/10.1016/j.tifs.2011.12.007
2. Yogesh K, Ahmad T, Manpreet G, Mangesh K, Das P (2013) Characteristics of chicken nuggets as affected by added fat and variable salt contents. J Food Sci Technol 50(1):191–196. https://doi.org/10.1007/s13197-012-0617-z PMID:24425908
3. Verma AK, Banerjee R, Sharma BD (2012) Quality of low fat chicken nuggets: effect of sodium chloride replacement and added chickpea (Cicerarietinun L.) hull flour. Asian-Aust J Anim Sci 25:291–298. https://doi.org/10.5713/ajas.2011.11263
4. Jiménez-Colmenero F, Carballo J, Cofrades S (2001) Healthier meat and meat products: their role as functional foods. Meat Sci 59:5–13. https://doi.org/10.1016/s0309-1740(01)00053-5
5. World Health Organization (WHO) (2023) Sodium reduction. Disponível em: https://www.who.int/news-room/fact-sheets/detail/salt-reduction. Accessed: 27 Nov 2024
6. Rodrigues I, Gonçalves LA, Carvalho FA et al. (2020) Understanding salt reduction in fat-reduced hot dog sausages: Network structure, emulsion stability and consumer acceptance. Food Sci Technol Int 26(2):123–131. https://doi.org/10.1177/1082013219872677
7. Barros JC, Rodrigues I, Pires MA, Gonçalves LA, Carvalho FAL, Trindade MA (2019a) Healthier chicken nuggets incorporated with chia (*Salvia hispanica* L.) flour and partial replacement of sodium chloride with calcium chloride. Emir J Food Agric 31(10):794–803. https://doi.org/10.9755/ejfa.2019.v31.i10.2021
8. Santos BA, Campagnol PCB, Fagundes MB, Wagner R, Pollonio MAR (2017) Adding blends of NaCl, KCl, and CaCl2 to low-sodium dry fermented sausages: Effects on lipid oxidation on curing process and shelf life. J Food Quality 2017:7085798. https://doi.org/10.1155/2017/7085798
9. Almeida MA, Villanueva NDM, Pinto JSS, Saldaña E, Contreras-Castillo CJ (2016) Sensory and physicochemical characteristics of low sodium salami. Scientia

Agrícola 73(4):347–355. https://doi.org/10.1590/0103-9016-2015-0096
10. Alrawashdeh H, Abu-Alruz K (2022) Development of high-fiber, low fat chicken nuggets. Int J Food Stud 11(2):354–373. https://doi.org/10.7455/ijfs/11.2.2022.a8
11. Doyle ME, Glass KA (2010) Sodium reduction and its effect on food safety, food quality, and human health. Compr Rev Food Sci Food Saf 9:44–56. https://doi.org/10.1111/j.1541-4337.2009.00096.x
12. Ruusunen M, Puolanne E (2005) Reducing sodium intake from meat products. Meat Sci 70:531–541. https://doi.org/10.1016/j.meatsci.2004.07.016
13. Verma AK, Banerjee R, Sharma BD (2015) Quality characteristics of low fat chicken nuggets: Effect of salt substitute blend and pea hull flour. J Food Sci Technol 52(4):2288–2295. https://doi.org/10.1007/s13197-013-1218-1 PMID:25829611
14. Barros JC, Gois TS, Pires MA, Rodrigues I, Trindade MA (2019b) Sodium reduction in enrobed restructured chicken nuggets through replacement of NaCl with CaCl2. J Food Sci Technol 56(8):3587–3596. https://doi.org/10.1007/s13197-019-03777-8 PMId:31413386
15. Segel IH (1976) Biochemical calculations: how to solve mathematical problems in general biochemistry, 2nd edn. Wiley, New York
16. Barbut S (2015) The science of poultry and meat processing. In: Battering and breading – production under HACCP. University of Guelph, Guelph, pp 14-1–14-50
17. Barros JC, Munekata PES, Pires MA, Rodrigues I, Andaloussi OS, Rodrigues CEC, Trindade MA (2018) Omega-3- and fibre-enriched chicken nuggetsby replacement of chicken skin with chia (*Salvia hispanica* L.) flour. LebensmittelWissenschaft + Technologie 90:283–289. https://doi.org/10.1016/j.lwt.2017.12.041

Chapter 8

Purple Sweet Potato (*Ipomea batata* L) Peel to Enhance the Self-Life of Beef Patties

Raquel Lucas-González, Rubén Domínguez-Valencia, Miriam Pateiro and José Manuel Lorenzo Rodriguez

Abstract

Purple sweet potato peel is a rich source of anthocyanins and exhibits strong antioxidant activity. By using emerging techniques, such as ultrasonic-assisted extraction coupled with a sonotrode to obtain bioactive compounds, and encapsulation methods to extend shelf life and stabilize these compounds, it is possible to produce a stable, natural colorant and antioxidant additive suitable for use in the meat industry. This development responds to consumer demand for clean-label products and environmentally respectful processes. Furthermore, the use of peel valorizes coproducts from the agro-food industry, thereby promoting a circular economy. This chapter presents a step-by-step methodology to obtain encapsulated anthocyanins from purple sweet potato peel and their subsequent application in beef burgers.

Keywords Sonotrode, Emergent technologies, Anthocyanins, Additives, Meat

1 Introduction

Meat and meat products are widely consumed worldwide. Their global demand is expected to increase in the next 30 years [1, 2]. Meat and meat products are fresh products susceptible to microbial deterioration and lipid and protein oxidation. Therefore, the main meat products present in the market contain additives, which preserve food security and organoleptic characteristics [3].

Natural additives have gained attention in the last decades, following the trend of green labels. Consumers are more conscious of health–food pairing reject manufactured food products with high ingredients and synthetic additives content [4, 5]. On the other hand, food industry tries to reduce their CO_2 feed in their manufacturing process and to promote circular economy. Thus, valorization of agro-food industry coproducts has an alternative to reduced waste and to obtain new sources of natural additives [6, 7].

José Manuel Lorenzo Rodriguez (ed.), *Meat Product Quality Improvement*, Methods and Protocols in Food Science,
https://doi.org/10.1007/978-1-0716-5460-6_8,

Anthocyanins are a kind of water-soluble polyphenol characterized by their red–blue–purple color and the ability to modify their color with pH changes. Currently, anthocyanins are considered as a colorant additive (E-163), mainly those obtained from grapes, black carrots, and red cabbages. Moreover, anthocyanins present also antioxidant activity, make then interesting to use as antioxidant additives [8].

Following the eco-friendly production of manufactured foods and additives, many emergent technologies have been explored to obtain natural additives from coproducts [9]. Purple sweet potato peels are a good source of anthocyanins. Due to the increase in the production of foods containing or made with purple sweet potato peel are present as a coproduct. An innovative and with lees environmental impact to obtained anthocyanins extracts with high antioxidant activity is extractions with sonotrode probe. Cavitation phenomena contribute to break cell wall and release pigments from vacuole to extractant medio in a short period time [10]. To obtain colorant and antioxidant additive from purple sweet potato peel, the whole process has three different steps: (1) Purple sweet potato peel obtention. The process involves the unit operations of brushing, peeling, dehydration, grinding, and sieving; (2) A solid-liquid extraction is performed using an ultrasonic probe that utilises cavitation forces; and (3) Encapsulation. The concentrated aqueous extract is dehydrated by spray-drying using at least one encapsulating agent. The encapsulated anthocyanins from purple sweet potatoes, with colourant and antioxidant properties, were used to increase the shelf life of beef burgers.

2 Materials

Purple sweet potato (*Ipomea batata*) and maltodextrin were bought to a local commercial, Ourense, Spain.

Ethanol 60% pH 3.0. Pure ethanol to food grade was mixed with water to obtain ethanol to 60%. For a 1 L preparation, 600 mL of ethanol was mixed with 400 mL to distilled water. After mixing ethanol and water, the pH of the solution was achieved to 3.0 using a solution of malic acid 1 M.

Sonication process be carried out in a ultrasonicador UP400St (Hielscher Ultrasonics GmbH, Germany) coupled to S24d22D probe. Rotavaporation process was carried out in a (Büchi Rotavapor R-200, BUCHI Ibérica S.L.U, Spain). For Spray drying, Mini Spray Dryer B-290 (BÜCHI Labortechnik GmbH) was employed.

3 Methods

The whole process to obtain a natural antioxidant and colorant from purple sweet potato peel and their application in beef burgers are described below (Fig. 1):

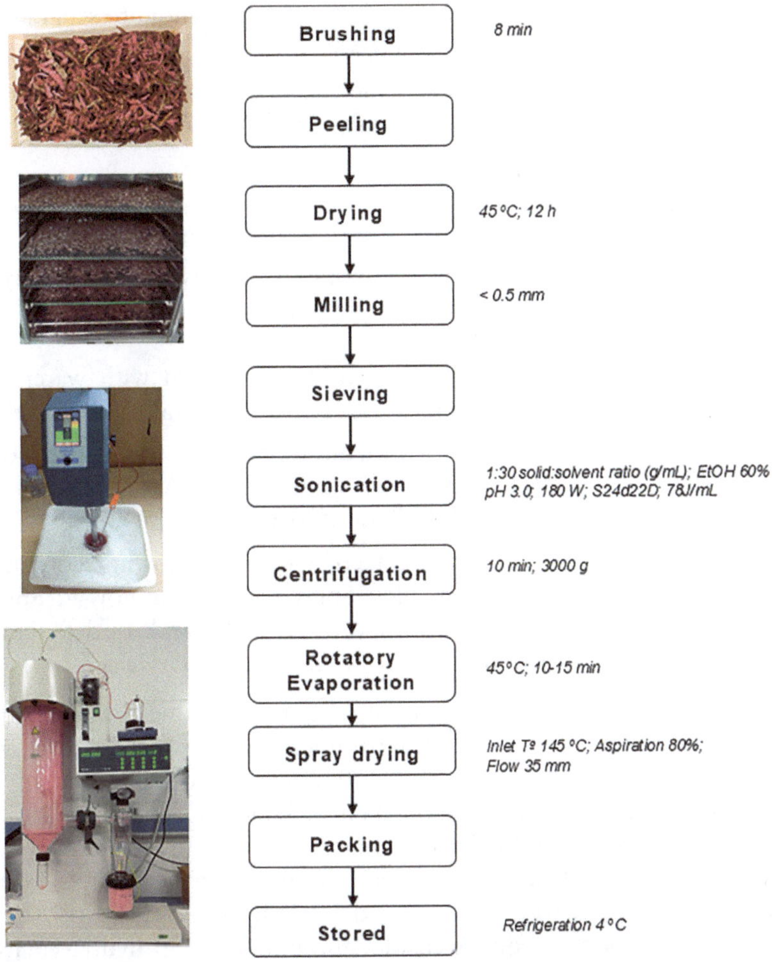

Fig. 1 Flow chart of the encapsulated anthocyanins from purple sweet potato peel additive

1. Brush purple sweet potato for 8 min in an automatic potato dry-cleaning machine.
2. Peel Purple sweet potatoes manually.
3. Dehydrate the peel in an air force oven to 12 h to 45°C.
4. Mill and sieve peel until achieve particle size < 1 mm.
5. Weigh 33.3 g of dehydrated purple peel in a 1.0 L glass and mix with 1.0 L of ethanol pH 3.0 (60%) (*see* **Note 1**).
6. Mix dehydrated peel and ethanol (60%; pH 3.0) in a magnetic agitator for 1 min.
7. Place mixed in an ice bath and introduced sonotrode (S24d22D) in a mixed (deep 45 cm).
8. Start ultrasound process in a ultrasonicador UP400St: applied 78 J/mL to 180 W during around 30 min in an ice bath (*see* **Note 2**).

9. Centrifugate all content of extraction, to separate liquid phase to solids to 3000*g* for 10 min.
10. In a 2 L spherical balloons, introduced around 250 mL of extractions and submitted to evaporation under vacuum (*see* **Note 3**)
11. The evaporation step was finish when around 20 mL of liquid are remaining into the balloon (*see* **Note 4**).
12. Repeat the rotavaporation process three times more, until all the ethanol is evaporated.

13. Concentrated extracts were diluted with water to achieve a concentration of 5 g/100 mL of total solids (*see* **Note 5**).
14. Mix extracts with maltodextrin (13.5 g/100 mL of extract) in a ultraturrax (T 25 digital ULTRA-TURRAX®; IKA Works Spain) to 10,000 rpm/30 min.
15. Dry spray the mixed with the following conditions: 145°C Temperature Inlet; Spray gas volume flow at 35 mm (rotameter); aspirator speed at 80%; and outlet temperature at 67°C–70°C. The evaporation capacity was 1 L/h (*see* **Note 6**).

After encapsulation, a light violet powder was obtained (Fig. 2). The physicochemical parameters, antioxidant activity and the main anthocyanins of the natural additive obtained from purple sweet potato peel can be observed in Table 1. Composed of 20 anthocyanins, of which the five major anthocyanins are peonidin-3-caffeoyl-p-hydroxybenzoyl sophoroside-5-glucoside, peonidin-3-hydroxybenzoyl sophoroside-5-glucoside, cyanidin-3-caffeoyl-p-hydroxybenzoyl sophoroside-glucoside, and cyanidin-3-p-hydroxybenzoyl sophoroside-5-glucoside, whose anthocyanin concentration was 2.4 g/kg (Table 1). Furthermore, it presented antioxidant power by four different methods of antioxidant activity, DPPH [DPPH

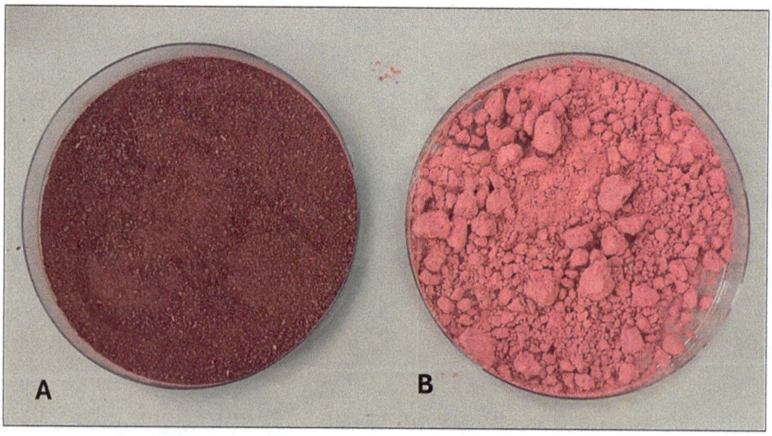

Fig. 2 (**A**) Purple sweet potato peel. (**B**) Encapsulated extract of anthocyanins from purple sweet potato peel

(2,2-diphenyl-1-picrylhydrazyl) free radical scavenging capacity], ABTS [ABTS (2,2'-Azino-bis (3-ethylbenzothiazoline-6-sulfonic acid) radical scavenging capacity], ORAC (oxygen radical absorption capacity), and FRAP (iron reducing antioxidant power assay).

The process of making beef burgers with encapsulated purple sweet potato peel extracts is described below. Furthermore, the flow diagram can be seen in Fig. 3.

1. Fresh beef must be used to make burgers.
2. Cut the beef into regular pieces (approximately 5×5 cubes) and remove remains of skin, bone, and cartilage.
3. Mince beef in an industrial mincer (plate of 8 mm).
4. Mix the minced beef with salt, water, and anthocyanin extracts (*see* **Note 7**).
5. Left the dough to rest for 15 min at 4°C.
6. Make 100 g balls with the meat dough.

Table 1
Antioxidant activity and color coordinates of the natural anthocyanin-based powder additive derived from the skin of purple sweet potato (*Ipomoea batata* L)

Anthocyanin profile	
Peonidin-3-caffeoyl-p-hydroxybenzoyl sophoroside-5-glucoside	1109.34±12.2
Peonidin-3-hydroxybenzoyl sophoroside-5-glucoside	301.44±3.01
Cyanidin-3-caffeoyl-p-hydroxybenzoyl sophoroside-glucoside	288.55±3.97
Peonidin-3- caffeoyl -feruloyl sophoroside-5-glucoside	160.27±7.11
Cyanidin-3-p-hydroxybenzoyl sophoroside-5-glucoside	134.07±2.82
Total, Anthocyanin content	2434.76±28.73
Antioxidant activity	
DPPH (mg trolox equivalent/g)	26.63±0.40
ABTS (mg ascorbic acid equivalent/g)	20.89±0.18
FRAP (μmol Fe^{+2}/g)	243.55±3.76
ORAC (mg trolox equivalent/g)	93.23±5.29
Color parameters	
L	64.45±1.15
a*	30.04±1.44
b*	−8.45±0.21
H°	105.72±0.46
C*	31.21±1.43

Data obtained to the [11]

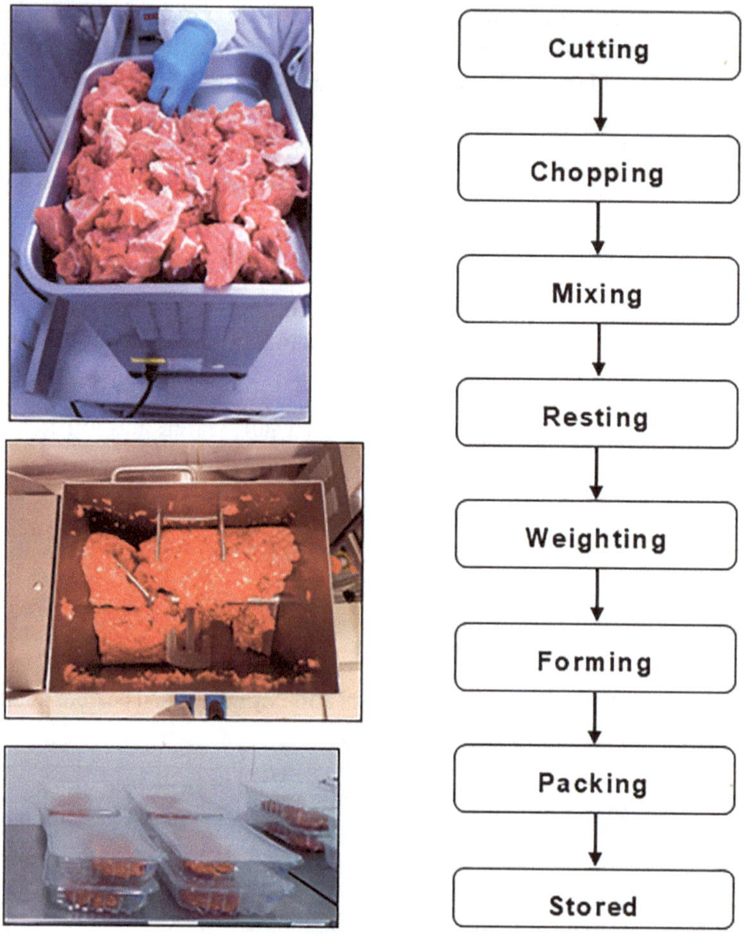

Fig. 3 Flow chart of beef burgers with encapsulated anthocyanins from purple sweet potato peel

7. Form the burgers with the help of a burger former.
8. Place two burgers in a plastic tray and store them under modified atmosphere.
9. Kept the burger under refrigeration.

To determine the colorant and antioxidant capacity of encapsulated anthocyanins from purple sweet potato peel, two additional burger formulations were made: one without additives and one with a synthetic additive. A shelf-life study was carried out over 9 days of storage (Fig. 4). Encapsulated anthocyanins from purple sweet potato maintained color stability throughout the storage period compared to burgers without additives. Furthermore, lipid oxidation values were lower than those observed in burgers without additives and in those with synthetic additives, as reported by Moraga-Babiano et al. [11]. These results indicate that encapsulated anthocyanins from purple sweet potato peel exhibit both colorant and antioxidant properties that can help preserve meat products.

Fig. 4 Image of the burgers at day 9 developed without additives, with natural additive (encapsulated anthocyanins from purple sweet potato peel) and with synthetic additive (sodium erythorbate)

4 Notes

1. When ethanol:water is mixed with purple sweet potato peel powder, superficial gelification can occur, blocking water diffusion into the rest of the powder. To avoid this phenomenon, add half the volume of the extractant slowly, stirring, if necessary, with a glass rod, which you can then clean with the rest of the extractant.
2. The ultrasonic process can be automated in two ways: by time and by specific energy. Search for the correct for you. You should choose the option that works best for your application. The most useful method is working with specific energy per volume, since this allows the process to be replicated when changing the sonotrode or the extraction volume. For example, if a final extraction of a 200 mL has a total energy input of 8000 Ws or J, the specific energy input per milliliter is calculated as follows:

$$\text{Specific energy input}\left(\frac{\text{J}}{\text{mL}}\right)=\frac{\text{Total energy input(J)}}{\text{Sample volume(mL)}}$$

   Therefore, the specific energy for the extraction is 40 J/mL. By multiplying this value by the new volume, it can adjust the required total energy input in the ultrasonic processor.
3. Rotavapor is a crucial step. This should be as quick as possible. To optimize the process, the balloon can only be filled one-fourth of its capacity. This allows for rapid boiling without the contents of the flask entering the coil. The bath temperature cannot overcome the temperature to the target compound is degraded. The difference between bath and coil temperature must be at least 20°C. Try to use an ice bath with recirculation to feed coil.
4. The final volume to achieve during the evaporation process depends on the initial concentration of the extraction mixture.

Furthermore, it is recommended not to dry the extract completely to prevent the overheating of bioactive compounds, facilitate product recovery, and conserve water. Nonetheless, care must be taken to ensure the complete evaporation of the organic phase, as a water-based solution is ultimately required for the subsequent spray drying process.

5. To know the real concentration, determined total solids to the extracts. For this purpose, weigh empty glass tubes. Then add 1.00 mL of extract and weight again. Left the glass with the sample 24 h or overnight in an oven a 105°C. After that weigh the tube with dehydrated extract. The total solid is calculated as follows:

$$\text{Total solids } (\%) \ \frac{\text{Weight tube with dehydrated sample } (g) - \text{emptytubeweight(g)}}{\text{extract volume} \quad (\text{mL})} \times 100$$

6. The Mini Spray Dryer B-290 does not allow you to control the outlet temperature. If it is too high before starting the process, around 80°C–90°C, run water through it until it drops to 67°C–70°C.

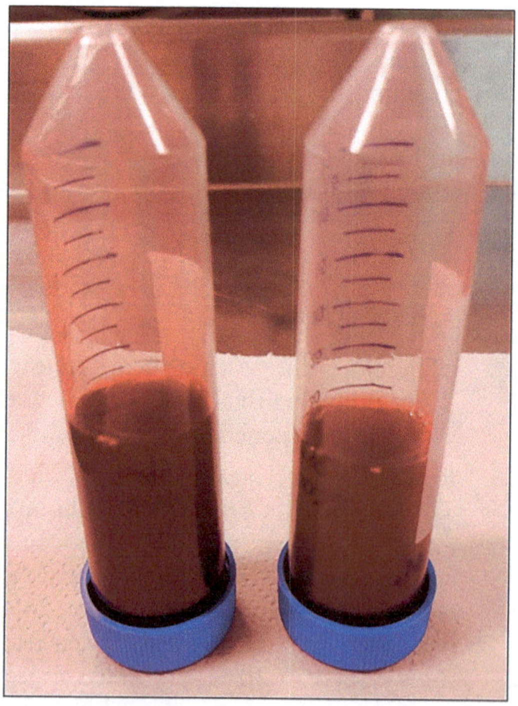

Fig. 5 Encapsulated anthocyanins from purple sweet potato peel diluted in one-sixth of the water added to beef burgers

7. The way to incorporate the process is as follows: layer to minced meat, layer to water and salt, layer to minced meat, and layer to water and salt. The encapsulated extract of anthocyanins from purple sweet peel must be completely dissolved in one-sixth of the total water used in a formulation. The best way to do it is in a tube and agitated vigorously. When it is completely dissolved, as in Fig. 5, it can be added to the meat dough.

Acknowledgments

Rubén Domínguez-Valencia, Miriam Pateiro, and José M. Lorenzo are members of the Competitive Reference Group "FunFood" (Functional Food), funded by GAIN (grant number IN607A2023/01). This study was supported by the Spanish Ministry of Science and Innovation project (PID2021-123628OB-C41-Agri-food coproducts as a source of bioactive compounds for the development of functional meat products), which is funded by MCIN/AEI/10.13039/501100011033/ and FEDER, UE.

References

1. Grasso S, Estévez M, Lorenzo JM, Pateiro M, Ponnampalam EN (2024) The utilization of agricultural by-products in processed meat products: effects on physicochemical, nutritional and sensory quality—invited review. Meat Sci 211:109451. https://doi.org/10.1016/j.meatsci.2024.109451
2. Siegrist M, Hartmann C (2023) Why alternative proteins will not disrupt the meat industry. Meat Sci 203:109223. https://doi.org/10.1016/j.meatsci.2023.109223
3. Bellucci ERB, Bis-Souza CV, Domínguez R, Bermúdez R, Barretto ACDS (2022) Addition of natural extracts with antioxidant function to preserve the quality of meat products. Biomolecules 12:1506. https://doi.org/10.3390/biom12101506
4. Grant KR, Gallardo RK, McCluskey JJ (2021) Consumer preferences for foods with clean labels and new food technologies. Agribusiness 37:764–781. https://doi.org/10.1002/agr.21705
5. Asioli D, Aschemann-Witzel J, Caputo V, Vecchio R, Annunziata A, Næs T, Varela P (2017) Making sense of the "clean label" trends: a review of consumer food choice behavior and discussion of industry implications. Food Res Int 99:58–71. https://doi.org/10.1016/j.foodres.2017.07.022
6. Hadidi M, Orellana-Palacios JC, Aghababaei F, Gonzalez-Serrano DJ, Moreno A, Lorenzo JM (2022) Plant by-product antioxidants: control of protein-lipid oxidation in meat and meat products. LWT 169:114003. https://doi.org/10.1016/j.lwt.2022.114003
7. Nieto G, Fernández-López J, Pérez-Álvarez JA, Peñalver R, Ros-Berruezo G, Viuda-Martos M (2021) Valorization of citrus co-products: recovery of bioactive compounds and application in meat and meat products. Plants 10:1069. https://doi.org/10.3390/plants10061069
8. Lucas-González R, Pateiro M, Domínguez-Valencia R, Carrillo C, Lorenzo JM (2025) Optimization of anthocyanin extraction from purple sweet potato peel (*Ipomea batata*) using sonotrode ultrasound-assisted extraction. Foods 14:2686. https://doi.org/10.3390/foods14152686
9. Basile G, De Luca L, Sorrentino G, Calabrese M, Esposito M, Pizzolongo F, Romano R (2024) Green technologies for extracting plant waste functional ingredients and new food formulation: a review. J Food Sci 89:8156–8174. https://doi.org/10.1111/1750-3841.17487
10. Martínez-Olivo AO, Durán-Castañeda AC, López-Cárdenas FG, de Rodríguez-Romero JJ, Sánchez-Burgos JA, Sáyago-Ayerd SG, Zamora-Gasga VM (2023) Advances in

ultrasound-assisted extraction of bioactive compounds (antioxidant compounds) from agrofood waste. In: Advanced technologies in wastewater treatment. Elsevier, pp 217–246
11. Moraga-Babiano L, Lucas-González R, Domínguez-Valencia R, Gaona-Ruiz M, Carrillo C, Echegaray N, Pateiro M, Lorenzo JM (2025) Encapsulated purple sweet potato peel extract as antioxidant and sustainable colourant to preserve the quality of beef burgers during the shelf life. Food Chem 487:144657. https://doi.org/10.1016/j.foodchem.2025.144657

Check for updates

Chapter 9

Addition of Xoconostle (*Opuntia joconostle*) Peel Extracts as an Antioxidant in Mexican Chorizo

Daliana Carbajal-Padilla, Esteban Arteaga-Cabrera and Julián A. Gómez-Salazar

Abstract

Artisanal red chorizo, an emblematic food of Mexican gastronomy, represents a significant source of animal-derived proteins and lipids. It is typically prepared with pork meat (70%–80%) and pork back fat (30%–20%) and is usually sold fresh, as it is not intentionally cured. However, a significant challenge during the preparation and storage of this product lies in the oxidation of lipids or proteins, which adversely affects its quality, shelf life, and nutritional profile. In this context, there is growing interest in enhancing these products by incorporating natural functional ingredients derived from plant materials native to Mexican regions, which not only help preserve desirable sensory characteristics but also offer potential health benefits. Mexico is recognized as a provider of various prickly pear species, including *xoconostle* (*Opuntia joconostle*). The peel of this fruit is a promising source of phenolic compounds and other antioxidants that may enhance the oxidative stability of meat products, while also addressing consumer demand for healthier, more sustainable, and clean-label foods. Within this framework, the present chapter introduces a strategy to enhance the functional and nutritional profile of Mexican red chorizo by using *xoconostle* peel extracts (*Opuntia joconostle*), a product of significant cultural relevance.

Keywords Functional artisanal chorizo, Xoconostle peel, Bioactive compounds, Oxidative stability

1 Introduction

Sausages are traditional foods that form an integral part of the diet in many European countries. In Mexico, the Spanish conquest introduced a wide variety of sausages, among which chorizo has become the most popular. Chorizo is a Hispanic meat product made from ground pork, pork fat, and salt, spices, and additives. In contrast, Mexican chorizo is a widely consumed sausage made with pork meat (70%–80%) and fat (30%–20%), characterized as a fresh meat product that is neither cured nor aged, and which also contains a blend of chili peppers typical of Mexican gastronomy [1–3]. Moreover, there is a broad variety of Mexican chorizos (fresh, fermented, and dry), with the incorporated ingredients varying according to the production

José Manuel Lorenzo Rodriguez (ed.), *Meat Product Quality Improvement*, Methods and Protocols in Food Science, https://doi.org/10.1007/978-1-0716-5460-6_9,

region and level of industrialization (artisanal, semi-industrial, and industrial) [2, 4]. In this context, the states of Hidalgo, Veracruz, Oaxaca, and the State of Mexico are the main producers of this meat product (Fig. 1). Particularly in Toluca, in the State of Mexico, the production of these meat products is highly popular and has a significant economic impact on the region, serving as a prime example of its importance in the country [4]. Table 1 presents the proximate composition of industrial and artisanal chorizos.

Despite the significant importance of this product in the country, critical issues frequently arise that affect its physicochemical, sensory, sanitary quality, and shelf life. These problems are mainly attributed to the intrinsic characteristics of Mexican chorizos. Fresh products with slight acidification, no curing process, and a certain percentage of added fat are highly susceptible to lipid oxidation [4]. Lipid oxidation is a serious concern in this meat product, as oxidized heme pigments can participate in lipoprotein reactions, leading to color changes and triggering oxidative reactions in the fresh product [6].

In recent years, consumers have shown increasing interest in healthy but also ethical, and sustainable foods. This trend has driven the development of plant-based additives incorporating functional

Fig. 1 Artisanal Mexican chorizos: Toluca (**a**), Hidalgo (**b**), Oaxaca (**c**), and Veracruz (**d**)

Table 1
Proximate composition of industrial and artisanal chorizos (g/100 g of fresh product)

Chorizo variety	Conuntry of origin	Moisture	Protein	Fat	Carbohydrates	Salt	pH	A_a
Industrial	España	49.5	16.7	29.5	1.5	1.9	N/A	N/A
Industrial	México	38.7	13.1	40.2	4.7	2.6	5.1	0.95
Green chorizo from Toluca	México	54.8	16.9	15.1	8.6	2	4.7	0.91
Mexican red chorizo	México	65.60	18.0	10.69	2.96	2.75	5.59	N/A

Data obtained from published scientific articles [1, 5]

ingredients, emphasizing the avoidance of synthetic additives whenever possible. Consumers seek products that offer health benefits, such as improved nutritional composition, while also demanding the use of naturally derived compounds as colorants, flavorings, or antioxidants [7].

For this reason, using natural antioxidants to mitigate the effects of oxidation reactions in meat products is considered a promising alternative strategy to synthetic antioxidants, without compromising the sensory quality of the final product.

The fruit and vegetable agro-industrial sector currently generates a large amount of by-products, accounting for up to 60% of total production. Therefore, the valorization of these by-products for the extraction of value-added ingredients, such as antioxidants, contributes significantly to the circular economy and environmental sustainability [8]. Mexico produces and consumes various prickly pear species, including *xoconostle* (*Opuntia joconostle*). However, the surplus generated after its processing results in agro-industrial by-products that can account for up to 50% of the total fruit weight, primarily consisting of the peel. Previous studies have demonstrated that the peel of this fruit is a rich source of polysaccharides, phenolic compounds, and betacyanins, all of high industrial value and with notable antioxidant potential [9, 10]. Hence, it is essential to extract the antioxidants present in the peel of *Opuntia joconostle*, to describe the stages of the extraction process, and to detail their incorporation into a meat product such as Mexican fresh chorizo.

Several studies have explored the incorporation of natural compounds in the production of chorizo, aiming to improve its quality, extend shelf life, and meet consumer demand for healthier and more natural products. For instance, adding natural extracts such as green tea, chestnut, or beer has proven effective in enhancing the oxidative stability of chorizo during storage, offering an alternative to synthetic antioxidants. Furthermore, plant extracts rich in phenolic compounds—such as rosemary, garlic, and paprika, among others—have been evaluated as natural alternatives to synthetic preservatives and antioxidants in chorizo. These have demonstrated notable antioxidant capacity and contributed to improved sensory quality of the meat product. These studies reflect a growing trend in the meat industry toward using natural and functional ingredients in chorizo production, addressing consumer concerns about health, sustainability, and food quality [3, 11, 12].

This chapter outlines the production stages of Mexican chorizo enriched with xoconostle peel extracts, which have a high antioxidant content, as well as the formulation and characteristics of the final product. Figure 1 shows the artisanal Mexican chorizos: Toluca (a), Hidalgo (b), Oaxaca (c), and Veracruz (d).

2 Materials

2.1 Extraction of Xoconostle Peel Extract

Fresh xoconostle fruits were used, exhibiting the typical morphological characteristics of varieties cultivated in semiarid regions, with predominantly green skin and partial reddish pigmentation. The fruits were selected at their physiological maturity stage, based on medium firmness, uniform size (60–80 g in weight, 4–6 cm in diameter), and the absence of visible defects such as spots, bruises, or decay (*see* **Note 1**).

2.2 Preparation of the Extraction Solvent

The solvent used is a 70% (v/v) hydroalcoholic solution, prepared by mixing food-grade ethanol (96%) with distilled water in a 70:30 ratio. The preparation is carried out in clean glass volumetric flasks, with manual shaking until a homogeneous mixture is obtained (Fig. 2) [13]. Figure 2 presents the extraction of ethanolic extract from xoconostle peel.

2.3 Preparation of Mexican Chorizo

According to the Mexican Official Standard NOM-145-SSA1-1995, chorizo is defined as a cured meat product made with minced meat not smaller than 2 mm. Its presentation may include maturation processes through microbial cultures or the addition of sugars, and it may or may not undergo cooking and/or smoking.

The common ratio used commercially for chorizo production is lean pork meat (70%–80%) and pork fat or bacon (30%–20%) (*see* **Note 2**) to obtain high-quality products.

[3]. The characteristics of Mexican chorizos vary depending on the region and variety (*see* **Note 3**). Distinctive attributes of the product include:

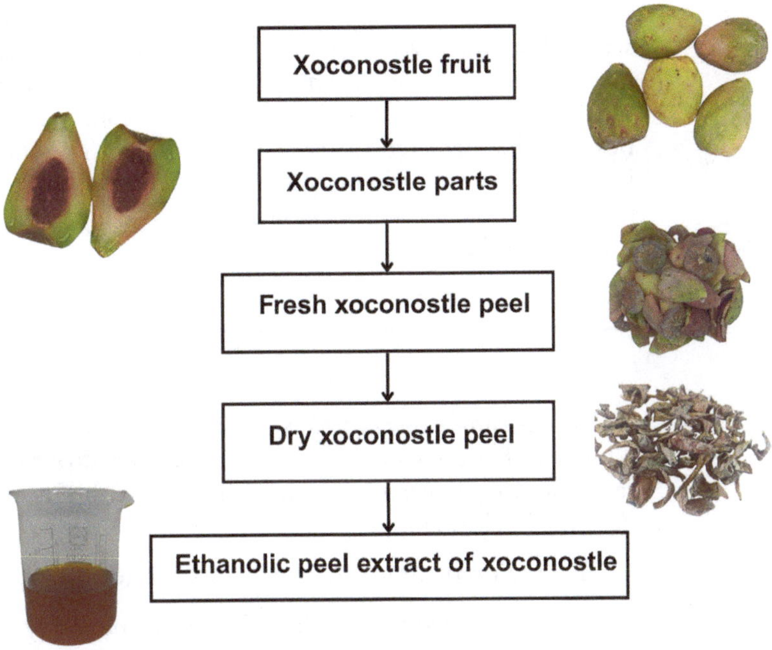

Fig. 2 Extraction of ethanolic extract from xoconostle peel

- Shape and External Appearance: Chorizo can be produced in various forms, including as a ball, whole sausage, or crumbled. When presented as a ball or whole sausage, the external appearance depends on the natural casing, which can be either natural (pork intestine) or artificial. The surface is smooth, without roughness, and glossy.
- Weight: The weight range varies depending on the chorizo presentation and the level of industrialization. Weight typically ranges from 0.5 to 2 kg.
- Consistency: The meat mass is firm and compact to the touch, perfectly adhered to the casing without gaps or separations.
- Color: Bright red is characteristic of the exterior casing or intestine. Inside, a more intense red color usually occurs due to lower oxidation of heme pigments, which have less contact with atmospheric oxygen. The quantity and variety of added chilies influence this attribute. Additionally, the addition of nitrates and nitrites changes the product.
- Aroma: Complex, spicy, with sweet and sour notes, resulting from pork meat, chilies, garlic, onion, vinegar, and various spices.
- Flavor: Intense, spicy, seasoned, with slightly acidic notes.
- Texture: Coarse, nonhomogeneous, granular, and firm to the touch.

For preparing artisanal red chorizo, pork-based raw materials and selected natural seasonings ensure high sensory and technological quality. Fresh pork leg is employed as the primary source of meat protein, while pork bacon is used as the fat source to provide juiciness and the characteristic flavor of the chorizo. The seasonings used include salt, curing salt, pepper, garlic, oregano, bay leaf, onion, and chili powder. Natural pork casing, previously hydrated, is required for stuffing the sausage.

3 Methods

The processing stages of Mexican chorizo, with and without the addition of xoconostle peel extract, are carried out using an artisanal method. The general steps for preparing this meat product are described below (Fig. 3).

Extraction of antioxidant-rich xoconostle peel extract

1. Xoconostle fruits are selected by discarding any anomalies in the raw material.

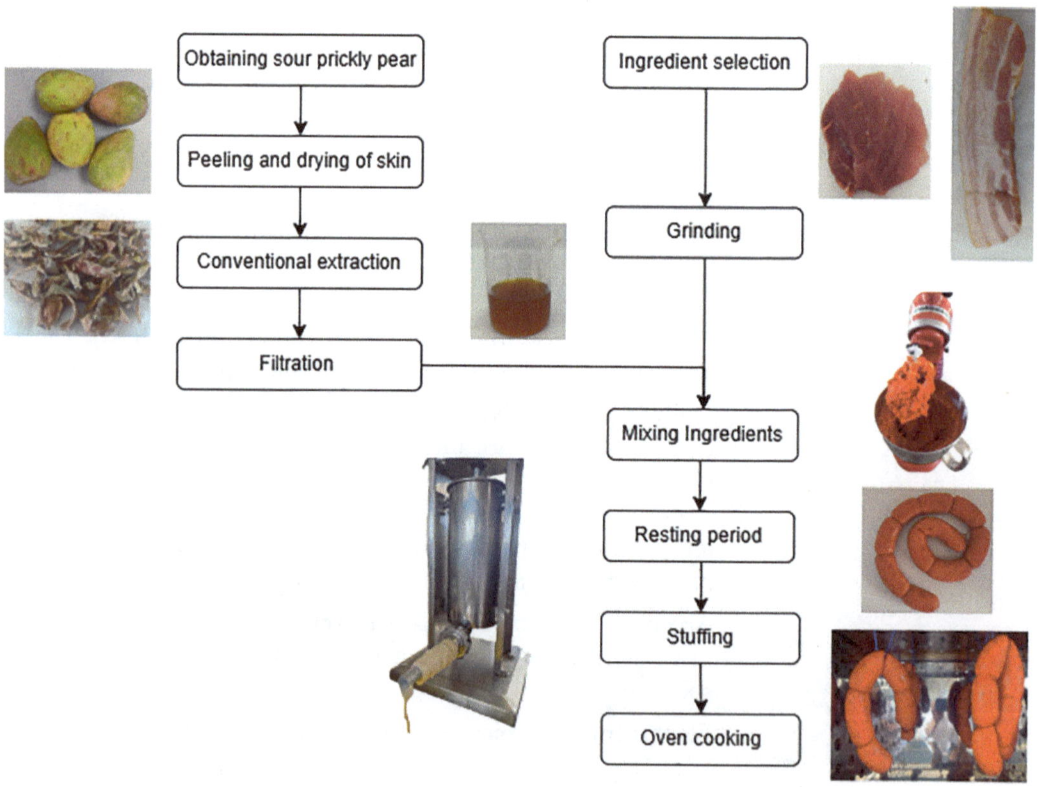

Fig. 3 Process for obtaining an artisanal chorizo enriched with *Opuntia joconostle* extract

2. The highest-quality fruits are washed and their peels removed. This by-product is spread out and dried at 60°C for 24 h. The fruit pulp is reserved for other preparations.
3. After drying, the by-product is ground, sieved, and subjected to hydroalcoholic extraction with ethanol (70% v/v) at 10:75 (sample to solvent) for 50 min at 60°C and 400 rpm.
4. Finally, filtration separates the extract from impurities and makes it ready to be used (*see* **Note 4**).

Artisanal Mexican Chorizo Preparation.

1. Pork meat and fat are selected with a pH ranging between 5.5 and 6.0, at a temperature of 4°C.
2. The animal-derived ingredients are cut into uniform pieces and ground using a 6–8 mm grinding plate.
3. Kneading occurs by mixing the ground meat with chillies, spices, additives, and salt until a homogeneous mass is obtained (approximately 8 min) with an even distribution of ingredients. At this stage, the xoconostle extract can be added before the rest of the ingredients to evenly distribute antioxidants within the meat (*see* **Note 5**).
4. Subsequently, the mixture rests for 6–12 h at 4°C and 85%–90% relative humidity.
5. After this step, stuffing is performed, typically using natural pork casing with a 32–34 mm diameter. Simultaneously, cotton twine is manually portioned the chorizo into characteristic 10 cm sections (*see* **Note 6**).
6. Finally, the product can be marketed fresh or heat-treated. A thermal treatment is applied in a convection oven at 80°C until a minimum internal temperature of 70°C is reached (*see* **Note 7**) [14].
7. The resulting chorizos can be commercialized using vacuum packaging, in bulk presentations, or as precooked products while preserving their sensory characteristics (*see* **Notes 8 and 9**).

Figure 3 shows the process for obtaining an artisanal chorizo enriched with *Opuntia joconostle* extract.

4 Notes

1. The xoconostle peel extract must be obtained from *fruits of the Opuntia genus* that exhibit predominantly green hues with slight reddish tones, which indicates adequate phenolic maturity. The fruits must be firm, show no signs of fermentation,

and be washed, peeled, and chopped before the extraction process.

2. The pork meat must come from healthy animals fit for human consumption, preferably young adult pigs. Cuts such as pork leg and pork fat (bacon) are recommended due to their technological functionality and sensory contribution.
3. To comply with Mexican regulations (NOM-086-SSA1-1994), the meat content must represent at least 70% of the formulation, and pork fat (bacon) should be included in proportions between 20% and 30%, depending on the desired texture and juiciness.
4. It is recommended that the extract be obtained in advance of the chorizo preparation. The extract must be stored under refrigeration (4°C) and used quickly to avoid losses in antioxidant activity or microbiological alterations.
5. Based on previous antioxidant capacity studies, the extract is incorporated into the chorizo formulation at a defined percentage. This addition is made when the ground meat is mixed with the seasonings.
6. The chorizos are stuffed into natural pork casings, pricked to eliminate air bubbles, and rest under refrigeration for 12–24 h to promote flavor integration.
7. The chorizos are cooked in hot-air convection ovens until reaching a minimum internal temperature of 70°C, ensuring the product's microbiological safety, especially against pathogens such as *Listeria monocytogenes.*
8. The shelf life of chorizo made with natural ingredients and stored under refrigeration (4°C–6°C) can be up to 21 days, provided that hygienic conditions are maintained and appropriate packaging is used [5].
9. The chorizo containing the xoconostle peel extract exhibited a slightly darker hue than the control. However, this change did not negatively affect the essential sensory characteristics of the product, such as the typical aroma and flavor of Mexican chorizo.

Acknowledgments

The authors thank SECHITI (Secretaria de Ciencias, Humanidades, Tecnología e Innovación de México) for the financial support for Daliana Carbajal-Padilla and Esteban Arteaga-Cabrera.

References

1. Aviles JCC, Garza CGG, Balandrano DDH et al (2019) El aceite esencial de orégano como sustituto del vinagre en la formulación de chorizo de puerco. Revista Chapingo Serie Zonas Áridas 18:11–25
2. González-Tenorio R, Totosaus A, Caro I, Mateo J (2013) Caracterización de propiedades químicas y fisicoquímicas de chorizos comercializados en la zona centro de México. Información Tecnológica 24:3–14
3. Perales-Jasso YJ, Gamez-Noyola SA, Aranda-Ruiz J et al (2018) Oregano powder substitution and shelf life in pork chorizo using Mexican oregano essential oil. Food Sci Nutr 6:1254–1260
4. Becerril Sánchez AL, Dublán García O, Domínguez-López A et al (2019) La calidad sanitaria del chorizo rojo tradicional que se comercializa en la ciudad de Toluca. Estado de México Rev Mex Cienc Pecu 10:172–185
5. Villalobos-Delgado LH, Caro I, Salvá BK et al (2022) Chorizo and chouriço de carne: varieties, composition, manufacturing process and shelf life. In: Pork: meat quality and processed meat products. CRC Press, pp 127–153
6. Maillard-Berdeja KV, Ponce-Alquicira E, Schettino-Bermúdez BS, Perez-Chabela ML (2022) Pomegranate (*Punica granatum* L.) peel flour as functional ingredient for chorizo: effect physicochemical and sensory characteristics of functional meat products. Acta Univ Cinbinesis, Ser E: Food Technol 26(1):33–42
7. McClements DJ, Grossmann L (2021) The science of plant-based foods: constructing next-generation meat, fish, milk, and egg analogs. Compr Rev Food Sci Food Saf 20:4049–4100
8. Zayed A, Farag MA (2020) Valorization, extraction optimization and technology advancements of artichoke biowastes: food and non-food applications. LWT 132:109883
9. Osorio-Esquivel O, Álvarez VB, Dorantes-Álvarez L, Giusti MM (2011) Phenolics, betacyanins and antioxidant activity in *Opuntia joconostle* fruits. Food Res Int 44:2160–2168
10. Amaya-Cruz DM, Pérez-Ramírez IF, Delgado-García J et al (2019) An integral profile of bioactive compounds and functional properties of prickly pear (*Opuntia ficus* indica L.) peel with different tonalities. Food Chem 278:568–578
11. Martínez L, Bastida P, Castillo J et al (2019) Green alternatives to synthetic antioxidants, antimicrobials, nitrates, and nitrites in clean label Spanish chorizo. Antioxidants 8:184
12. Pateiro M, Bermúdez R, Lorenzo JM, Franco D (2015) Effect of addition of natural antioxidants on the shelf-life of "Chorizo", a Spanish dry-cured sausage. Antioxidants 4:42–67
13. Gómez-Salazar JA, Patlán-González J, Sosa-Morales ME et al (2022) Multi-objective optimization of sustainable red prickly pear (*Opuntia streptacantha*) peel drying and biocompounds extraction using a hybrid stochastic algorithm. Food Bioprod Process 132:155–166
14. Martínez E, Pardo JE, Álvarez-Ortí M et al (2023) Substitution of pork fat by emulsified seed oils in fresh deer sausage ("Chorizo") and its impact on the physical, nutritional, and sensory properties. Foods 12:828

Chapter 10

Beef Patties Elaborated with Natural Antioxidants Derived from Encapsulated Kiwi Peel Extracts

Rubén Agregán, Noemí Echegaray, Raquel Lucas-González, Rubén Domínguez-Valencia, Mirian Pateiro and José Manuel Lorenzo Rodriguez

Abstract

Traditional beef patty recipes often incorporate synthetic additives in their formulation, which are attributed to long-term harmful effects on health. Therefore, plant-derived compounds are being investigated to replace synthetic additives in preventing oxidation phenomena during food product storage. The previously tested antioxidant and antimicrobial activities of kiwi peels allow their addition to highly perishable food products such as beef patties to naturally extend their shelf-life. This chapter presents the use of this plant residue as a natural preservative in the process of making a beef patty, which is proposed as an innovative recipe aimed at incorporating a rich-antioxidant waste material for the development of a value-added meat product. In addition, well-developed extraction and encapsulation processes for kiwi peels are suggested in order to enhance and protect its high biological properties.

Keywords Plant-derived extract, Waste organic material, Ultrasound-assisted extraction, Spray drying, Lipid oxidation, Bioactive compound

1 Introduction

The growing demand in the food market for products that meet the current consumer preferences, including the use of natural ingredients and longer shelf-life, has led to a gradual abandonment of synthetic additives potentially harmful to health, since, if used incorrectly, they could cause toxicity or increase the risk of cancer [1]. Plant-derived compounds have been assessed as replacers for those of synthetic origin, thanks to their recently reported antioxidant and antimicrobial activities. Food products added with known natural chemicals are perceived by consumers as healthier, simpler, and free of toxics, which are called "clean label" [2].

Fruit peels are bioactive-rich-organic materials frequently discarded in many food processing industries, constituting up to half of the waste generated by the commercial sector and households.

José Manuel Lorenzo Rodriguez (ed.), *Meat Product Quality Improvement*, Methods and Protocols in Food Science, https://doi.org/10.1007/978-1-0716-5460-6_10,

These residues contain abundant amounts of biologically active substances, such as phenolic compounds, proteins, lipids, fiber, and polysaccharides, making them suitable for use as nutraceuticals [3]. Kiwifruit (*Actinidia deliciosa*) processing industries produce large amounts of by-products, including peels, among others, such as roots or pulp, presenting all high levels of bioactive components for use as functional ingredients [4]. Both antioxidant and antimicrobial properties were examined in the kiwi peel, reporting a strong ability to scavenge reactive oxygen species (ROS), supported by the presence of carotenoids such as lutein, chlorophyll, ascorbic acid, and phenolic compounds, including phenolic acids, such as protocatechuic, chlorogenic, protocatechuic, and caffeic acids, and flavonoids, such as rutin and quercetin. Kiwi peel residue also displays inhibitory activity against foodborne pathogenic bacteria, such as *Staphylococcus aureus*, *Escherichia coli*, *Pseudomonas aeruginosa*, and *Listeria monocytogenes* [5–7]. These outcomes underline the potential of kiwi peel as a preservative ingredient in innovative food formulations.

The suitability of kiwi peels as natural shelf-life extenders in different meat matrices, such as sausages and beef burgers, was evaluated, revealing significant antioxidant activities during storage, which helped preserve the quality of the product for longer [8, 9]. The inhibition of bacterial growth when producing beef burgers with added kiwi peels was also highlighted [10]. The biological impact of this organic residue and others in complex food matrices is conditioned by the recovery method chosen to obtain the extract. In this context, emerging green extraction techniques are being developed and refined in order to improve processing efficiency, reduce energy consumption and losses, and ultimately better manage production costs [4], allowing for increased sustainability in the production chain. Among the techniques currently used, the performance of ultrasound-assisted technology (UAE) stands out owing to the permeability induced in plant cells, facilitating the release of bioactive compounds without using high temperatures and long processing times. However, the extracts obtained may be susceptible to oxidation phenomena, negatively impacting the sensory characteristics of the added products [9]. Encapsulation of plant-based extracts can mitigate certain unpleasant organoleptic properties, preventing the integration of weird flavors and odors into food. Concurrently, this technique can protect bioactives from aggressive external conditions (e.g., high temperatures, light exposure, and oxidative phenomena), increasing their stability and helping to control their release into the food matrix, which ultimately improves the final quality of the product [11].

This chapter describes in detail a comprehensive protocol for making beef patties, using encapsulated kiwi peel extracted by the UAE as a natural antioxidant. Beef patties are a widely consumed and appreciated processed meat worldwide [12], but they are prone to sensory and nutritional losses due to their particular preparation method. The muscle grinding increases the surface area in contact with the air by producing smaller particles, causing rancidity and exposing the lipid membranes to metal oxidation catalysts [13]. Therefore, the use of antioxidants is crucial to maintain oxidative stability during refrigerated storage. In this sense, encapsulated kiwi peel extract (KPE) can significantly delay the oxidation of lipids and proteins, extending the shelf-life of the meat product. The following paragraphs provide a detailed step-by-step procedure for obtaining these functional ingredients and producing the beef patties.

2 Materials

2.1 Ingredients

Only a few range of ingredients can be used to make beef patties, as these meat products do not require complex mixtures. Traditional recipes are based on beef, the main ingredient by weight, which represents approximately 90% of the dough for this recipe. Commercial preparations commonly use lower proportions of meat, mixed with higher amounts of fat. This protocol aims to reduce the fat content by adding only 5% in order to achieve a low-fat meat product. The addition of 7.5% of water to the meat dough is recommended in order to facilitate the formation of the patties and also to properly disperse the ingredients inside. It is suggested to use a small percentage of salt (NaCl) (1% in this recipe) to provide better flavor and texture, also promoting the water and fat retention process [14]. Additional ingredients, such as seasonings, industrial preservatives (e.g., nitrites, sulfate, or ascorbic acid), and binders and texturizers (e.g., egg and milk), can be used in commercial preparations, but they are discarded in this recipe to avoid an overloaded food matrix and the use of synthetic compounds. Instead, the recommendation is to use an encapsulated KPE previously obtained with UAE to prevent oxidative reactions and prolong shelf-life. Table 1 shows all ingredients used and their ratio.

Table 1
Ingredients and their proportions by weight used for the preparation of beef patties

Ingredients	Proportion (% by weight)
Lean meat	86.5
Fat	5
Water	7.5
Salt (NaCl)	1
Encapsulated kiwi peel extract (KPE)	0.5

The meat for this preparation protocol can be obtained from the loin of beef. In this case, the fat was not added separately, but was part of the meat, although the process of mixing is also acceptable and common. On the other hand, kiwi peels can be obtained manually or by mechanical separation, preferably from ripe kiwis, taking care to discard all the pulp. For the development of this production protocol, kiwis from the company Kiwi Atlántico S.A., based in Pontevedra, Spain, were used.

2.2 Machinery Used During Product Development (see Notes 1 and 2)

For a successful and adequate finish product, optimal machines are needed, along with their required accessories. Table 2 presents the list of industrial and laboratory machinery for the manufacture of beef patties added with encapsulated KPE. For the preparation of the beef patties, a series of items were carefully selected (Figs. 1 and 2), which are described below:

- Mincer: the first step in the process of making a beef patty is to properly grind the raw material, obtaining fine particles for the formation of a suitable dough. To successfully accomplish this target, it is essential to reach a certain particle size. Therefore, it is suggested to utilize a plate with an 8 mm diameter. This accessory will be placed on a refrigerated meat mincer (La Minerva, Bologna, IT, Italy). Other brands that guarantee the same particle size are also admissible.
- Burger press: once all ingredients are mixed, the meat dough is shaped using a burger press (A-2000, Gaser, Girona, Spain). Other brands can be employed, ensuring an adequate patty formation. This protocol recommends press machines capable of producing patties with measurements of 10 cm in diameter and 1 cm in height.

Table 2

List of industrial and laboratory machinery for the manufacture of beef patties, added with encapsulated kiwi peel extract (KPE) (*see* Notes 1 and 2)

Machinery	Model and brand
Mincer	La Minerva, Bologna, IT, Italy.
Burger press	A-2000, Gaser, Girona, Spain.
Heat sealer	A LARI3/Pn T-VG-R-SKIN, Ca.Ve.Co., Palazzolo, Italy.
Gas injector	Viduca, Alicante, Spain.
Ultrasound equipment	Hielscher UP400St, Teltow Germany.
Homogenizer	IKA®-Werke GmbH & Co. KG, Staufen, Germany.
Encapsulator equipment	Mini Spray Dryer B-290, BUCHI Ibérica S.L.U, Barcelona, Spain.

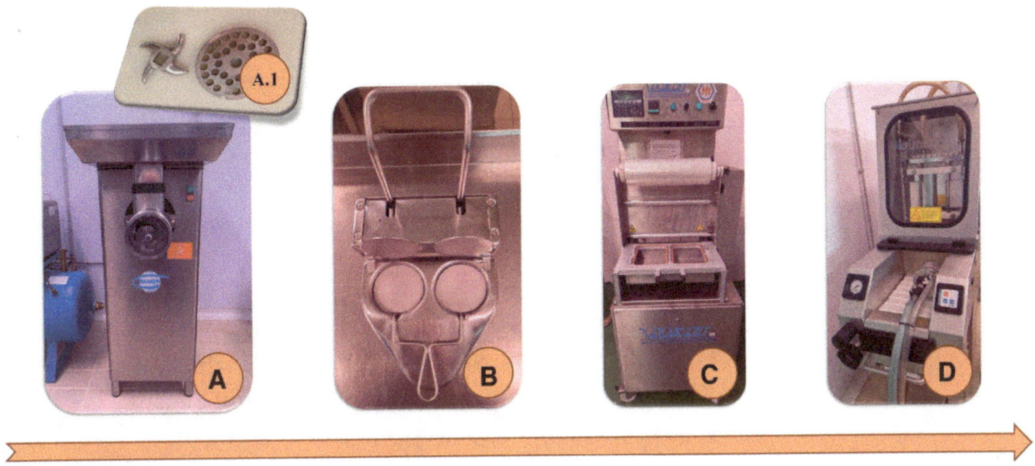

Fig. 1 Pictures of the different machines and accessories utilized in the production of beef patties. (**A**) mincer; (**A.1**) 8 mm plate; (**B**) burger press; (**C**) heat sealer; (**D**) gas injector. The machines are placed in order of use for the elaboration of the patties, from left to right

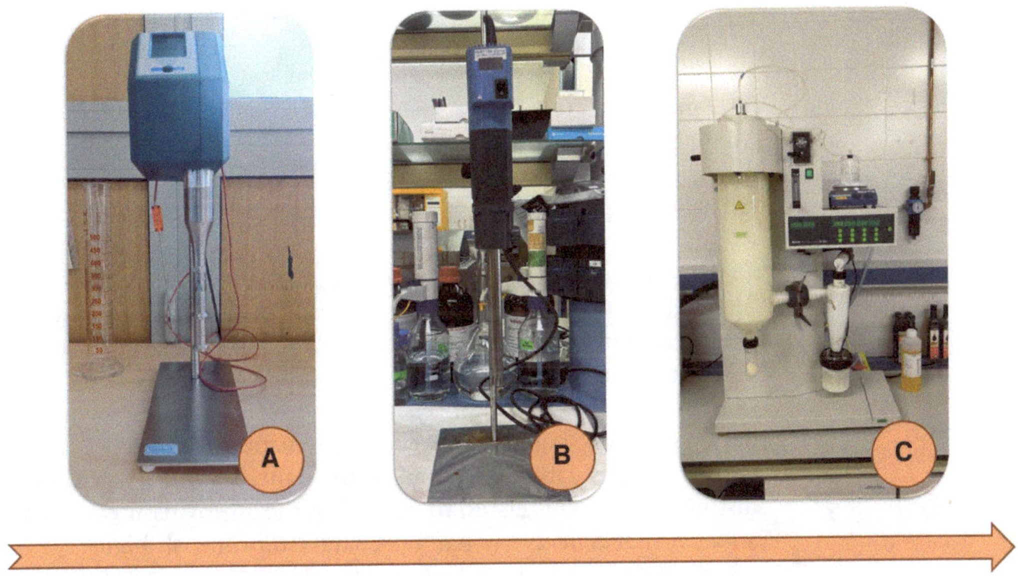

Fig. 2 Pictures of the equipment used in obtaining the encapsulated kiwi peel extract (KPE). (**A**) ultrasound equipment; (**B**) ultra-Turrax; (**C**) spray dryer. The different pieces of equipment are placed in order of use for the production of the extract, from left to right

- Heat sealer: the patties must be stored in a modified atmosphere, so they should be packaged using a sealer machine with a permeable cover to allow the passage of gases. A LARI3/Pn T-VG-R-SKIN heat sealer (Ca.Ve.Co., Palazzolo, Italy) is recommended, but other brands are susceptible to be used.

- Gas injector: O_2 and CO_2 gases must be added to the trays through the permeable covers using a gas injector from the Viduca brand (Alicante, Spain), although other brands are possible.
- Ultrasound equipment: the extract of previously chopped kiwi peels is obtained by a Hielscher UP400St ultrasonic device (Teltow, Germany) (optional brand), allowing the recovery of polyphenols under an ethanolic solution.
- Homogenizer: before spray drying, it is recommended to use an IKA T25 digital ultra-Turrax homogenizer (IKA®-Werke GmbH & Co. KG, Staufen, Germany) to thoroughly mix the KPE and the maltodextrin, the encapsulating agent.
- Encapsulator equipment: the concentrated KPE, together with maltodextrine, is transferred to a laboratory-sized spray dryer (Mini Spray Dryer B-290; BUCHI Ibérica S.L.U, Barcelona, Spain) to form the encapsulated components that will serve as an antioxidant ingredient in the beef patties.

3 Methods

The production of beef patties enhanced with encapsulated KPE involves multiple and specific complex stages. The novelty of the recipe lies in obtaining a kiwi peel extract and its subsequent encapsulation, requiring several delicate and sensitive steps, crucial to achieving the final goal of replacing synthetic antioxidants with another natural, value-added substance of plant origin. On the other hand, the elaboration of the patties also requires following a strict production protocol, paying attention to the quality of meat used and particularities of each stage, from the correct grinding of the meat to the storage conditions. The following paragraphs describe in detail all the stages necessary for the manufacture of beef patties with an encapsulated KPE as an antioxidant. Two main sections are distinguished: one related to obtaining the antioxidant extract and another that reflects the patty preparation protocol. Figures 3 and 4 schematically represent both protocols.

3.1 Obtaining the Encapsulated Kiwi Peel Extract (KPE)

3.1.1 Recovery of Kiwi Peel Antioxidants

The first step is to peel the kiwis carefully, removing as much pulp as possible. The process can be made easier by scalding the fruit beforehand. The peels are then washed with tap water and dried in an oven at 45°C for 24–36 h. Finally, they are ground into powder and, if not used the same day, should be vacuum-sealed and stored at freezing temperature.

The kiwi peel powder obtained is mixed with 60% ethanol, which acts as an extraction solvent, and the mixture is undergone to an ultrasonic process that allows the extraction of polyphenols and

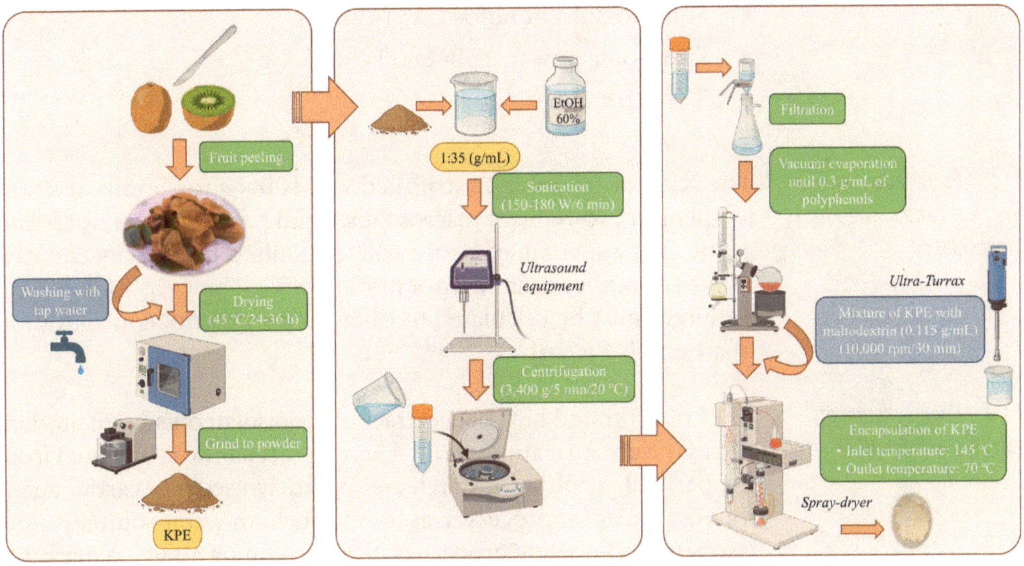

Fig. 3 Diagram of the different steps for the encapsulation process of kiwi peel extract (KPE)

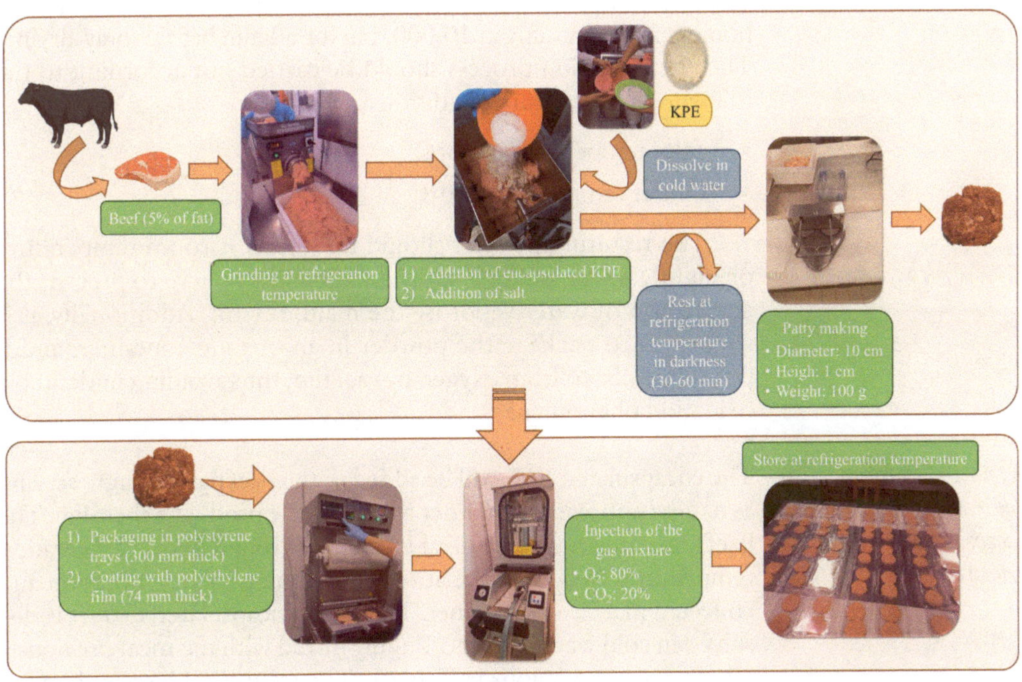

Fig. 4 Diagram of the different steps for the production of beef patties with encapsulated kiwi peel extract (KPE)

other antioxidant compounds. This step should be carried out under the following operating parameters:

- Solid-to-solvent ratio: 1:35 (g/mL)
- Ultrasonic power: 150–180 W
- Time: 6 min

3.1.2 Preparation of Kiwi Peel Extract (KPE) for Encapsulation

The KPE obtained is centrifuged at 3400 × *g* for 5 min at room temperature to remove macroscopic solid parts, such as fibers and pectin. The supernatant is then collected, filtered, and concentrated using a rotary vacuum evaporator at 45°C. The evaporation time required must be calculated to obtain a polyphenol concentration of around 0.3 g/mL.

3.1.3 Encapsulation of Kiwi Peel Extract (KPE)

The concentrated kiwi peel extract is encapsulated using maltodextrin as the encapsulating agent. This polysaccharide is obtained from the partial hydrolysis of starch and is widely used as a carrier agent for encapsulation processes, as it exhibits high water solubility, low viscosity, free-flowing properties, flavor-blending ability, remarkable binding properties, and effectively protects bioactives from oxidation phenomena [15, 16]. To achieve a proper encapsulation, the maltodextrin must be thoroughly mixed with the KPE (11.5 g/100 mL of extract). This step is best accomplished employing an ultra-Turrax homogenizer, running at 10,000 rpm for 30 min before spray-drying. The encapsulation process should be carried out according to the following temperature conditions:

- Inlet temperature: 145°C
- Outlet temperature: 70°C

The resulting powder should be stored at room temperature or slightly lower and in a low-humidity atmosphere to prevent the absorption of water vapor by the maltodextrin. Additionally, it is suggested to package the powder in an opaque container, and if possible, isolated from oxygen by vacuum, thus avoiding undesirable oxidation phenomena.

*3.1.4 Elaboration of Beef Patties with Added Encapsulated Kiwi Peel Extract (KPE) (see **Note 3**)*

The encapsulated KPE will be added to the beef patty dough, serving as natural antioxidant in order to extend the product's shelf-life. The beef loin with around 5% of fat is carefully ground under refrigeration temperature to avoid premature microbiological deterioration and collected in a clean container. Then, the encapsulated extract is dissolved in cold water and thoroughly mixed with the meat previously salted. See proportions in Table 1. The meat dough should be left to rest at refrigeration temperature and in darkness for between 30 and 60 min to facilitate the subsequent patty making, as it increases the consistency of the dough. The patties are shaped manually using a burger press, obtaining pieces about 10 cm in diameter and 1 cm high. The approximate weight per unit is 100 g. The next step in the patty making is capital for its proper preservation. Any variation or mistake at this point of elaboration could prematurely spoil the product and/or reduce its appeal, resulting in an unpleasant color and abnormal tastes, flavors, and odors. The patties must be packaged in polystyrene trays 300 mm thick and covered with a polyethylene film of 74 mm, with a permeability of less than 2 mL/(m^2bar/day). Then, a gas mixture consisting of 80% O_2 and 20% CO_2 is injected through the cover film. Finally, the patties must be stored refrigerated until their eventual retailment to prolong shelf-life, preventing premature oxidation and microbiological deterioration.

4 Notes

1. The models of the machines mentioned in this chapter are provided for guidance purposes only. The model may vary depending on the quantity of product to be manufactured. Therefore, the final decision regarding size and model is left to the reader's criteria.
2. The trademarks of the machines used in this chapter were mentioned simply as indicative examples, and their use is not mandatory. They were chosen based on the positive experience derived from their use by the authors.
3. The production of beef patties must be carried out entirely at refrigeration temperature to prevent premature microbiological deterioration.

Acknowledgments

Noemí Echegaray and Rubén Agregán acknowledge GAIN (Axencia Galega de Innovación) for their postdoctoral grants (grant numbers IN606C-2025/001 and IN606C-2025/002, respectively). Rubén Agregán, Noemi Echegaray, Rubén Domínguez-Valencia, Mirian Pateiro, and José M. Lorenzo are members of the Competitive Reference Group "FunFood" (Functional Food), funded by GAIN (grant number IN607A2023/01). This study was supported by the Spanish Ministry of Science and Innovation project (PID2021-123628OB-C41-Agri-food coproducts as a source of bioactive compounds for the development of functional meat products), which is funded by MCIN/AEI/10.13039/501100011033/ and FEDER, UE.

References

1. Novais C, Molina AK, Abreu RMV, Santo-Buelga C, Ferreira ICFR, Pereira C, Barros L (2022) Natural food colorants and preservatives: a review, a demand, and a challenge. J Agric Food Chem 70(9):2789–2805. https://doi.org/10.1021/acs.jafc.1c07533
2. Sbardelotto PRR, Balbinot-Alfaro E, da Rocha M, Alfaro AT (2023) Natural alternatives for processed meat: legislation, markets, consumers, opportunities and challenges. Crit Rev Food Sci Nutr 63(30):10303–10318. https://doi.org/10.1080/10408398.2022.2081664
3. Rather JA, Akhter N, Ayaz Q, Mir SA, Singh A, Goksen G, Majid D, Makroo HA, Dar BN (2023) Fruit peel valorization, phytochemical profile, biological activity, and applications in food and packaging industries: comprehensive review. Curr Food Sci Technol Rep 1(2):63–79. https://doi.org/10.1007/s43555-023-00007-3
4. Li K, Liu L, McClements DJ, Liu Z, Liu X, Liu F (2024) A review of the bioactive compounds of kiwifruit: bioactivity, extraction, processing and challenges. Food Rev Int 40(3):996–1027. https://doi.org/10.1080/87559129.2023.2212033
5. Alim A, Li T, Nisar T, Ren D, Zhai X, Pang Y, Yang X (2019) Antioxidant, antimicrobial, and antiproliferative activity-based comparative study of peel and flesh polyphenols from *Actinidia chinensis*. Food Nutr Res 63:1577. https://doi.org/10.29219/fnr.v63.1577
6. Wang Y, Li L, Liu H, Zhao T, Meng C, Liu Z, Liu X (2018) Bioactive compounds and *in vitro* antioxidant activities of peel, flesh and seed powder of kiwi fruit. Int J Food Sci Technol 53(9):2239–2245. https://doi.org/10.1111/ijfs.13812
7. Zeinab AS, Ahmed MA-E, Alaa AG, Abou-Elella F, Hanan FA, Mohsen SA, Habiba AA (2018) Active constituents of kiwi (*Actinidia deliciosa* Planch) peels and their biological activities as antioxidant, antimicrobial and anticancer. Res J Chem Environ 22(9):52–59
8. Boateng EF, Yang Z, Zhang W (2022) Effects of kiwifruit peel extract and its antioxidant potential on the quality characteristics of beef sausage. Antioxidants 11(8):1441. https://doi.org/10.3390/antiox11081441
9. Sirini N, Echegaray N, Moraga-Babiano L, Lucas-González R, Domínguez-Valencia R, Pateiro M, Lorenzo JM (2025) Application of encapsulated kiwifruit peel extract as a sustainable antioxidant in refrigerated beef patties. Food Biosci 64:105889. https://doi.org/10.1016/j.fbio.2025.105889
10. El-Nassag DE, Refaat WA (2017) Antioxidant and antibacterial activities of beef burger by using kiwi fruits peel powder. J Home Econ 27(1):31–53
11. Domínguez R, Pateiro M, Munekata PES, McClements DJ, Lorenzo JM (2021) Encapsulation of bioactive phytochemicals in plant-based matrices and application as additives in meat and meat products. Molecules 26(13):3984. https://doi.org/10.3390/molecules26133984
12. Agregán R, Barba FJ, Gavahian M, Franco D, Khaneghah AM, Carballo J, Ferreira IC, da Silva Barretto AC, Lorenzo JM (2019) *Fucus vesiculosus* extracts as natural antioxidants for improvement of physicochemical properties and shelf life of pork patties formulated with oleogels. J Sci Food Agric 99(10):4561–4570. https://doi.org/10.1002/jsfa.9694
13. Zahid A, Seo J-K, Park J-Y, Jeong J-Y, Jin S-K, Park T-S, Yang H-S (2018) The effects of natural antioxidants on protein oxidation, lipid oxidation, color, and sensory attributes of beef patties during cold storage at 4°C. Kor J Food Sci of Anim Res 38(5):1029–1042. https://doi.org/10.5851/kosfa.2018.e36
14. İncesu M, Akköse A (2024) Effect of different chloride salts on chicken burger quality. Int Food Res J 31(2):331–340. https://doi.org/10.47836/ifrj.31.2.05
15. Ahmadi F, Suleria HAR, Dunshea FR (2025) Physicochemical characterization, storage stability behavior, and intestinal bioaccessibility of clove extract encapsulated using varying combinations of gum Arabic and maltodextrin. Foods 14(2):237. https://doi.org/10.3390/foods14020237
16. Sahiner M, Gungor B, Silan C, Demirci S, Erdogan H, Ayyala RS, Sahiner N (2024) Synthesis and characterization of poly(maltodextrin) microgel from maltodextrin as drug delivery system. React Funct Polym 195:105826. https://doi.org/10.1016/j.reactfunctpolym.2024.105826

Chapter 11

Elderberry (*Sambucus nigra* L.) as a Natural Colorant to Extent the Beef Burger Shelf Life

Rubén Domínguez-Valencia, Mirian Pateiro, Aurora Cittadini, Rubén Agregán, Noemí Echegaray and José Manuel Lorenzo Rodriguez

Abstract

Elderberries are a largely underutilized fruit within the food industry, despite possessing excellent antioxidant and coloring properties. The plant is widely distributed across the globe and exhibits high resistance to adverse environmental conditions. While the fruit is commonly used in products such as beverages, jellies, yogurts, or ice creams, it is rarely found in other food categories. However, its potential application in the meat industry is immense, as it would serve a dual function (antioxidant and colorant) to replace synthetic additives, which are increasingly rejected by consumers demanding natural products. Therefore, it is crucial to clearly and concisely define the extraction, processing, and stabilization procedures for elderberry bioactive compounds, as there is currently no standardized protocol, and the varying conditions used across studies make it difficult to compare results. Aiming to address these limitations, this chapter provides a clear, comprehensive, and complete vision of a procedure for the processing, extraction, and encapsulation of elderberry bioactive compounds to obtain a natural additive. To this end, all stages are described in detail with specific conditions to ensure replicability by other researchers and scalability for industrial processes. Finally, the effect of this additive on a meat product (beef burger) is presented to demonstrate the benefits of using encapsulated extract over synthetic additives, highlighting the significant potential this natural ingredient holds for the meat industry.

Keywords Anthocyanins, Natural colorant, Natural additive, Antioxidants, Meat products, Elderberry, Cyanidin-3-O-sambubioside, Cyanidin-3-O-glucoside

1 Introduction

Elderberry (*Sambucus nigra* L.) is a robust shrub with great climatic adaptability, widely distributed throughout Europe [1, 2] and traditionally marketed in the form of juices, jams, as well as supplements or in beverages, jellies, ice creams, and yogurts [3, 4]. Despite being one of the plant sources with the highest concentrations of anthocyanins and polyphenols—surpassing other berries in antioxidant capacity [5]—it remains a remarkably underutilized resource.

José Manuel Lorenzo Rodriguez (ed.), *Meat Product Quality Improvement*, Methods and Protocols in Food Science, https://doi.org/10.1007/978-1-0716-5460-6_11,

This underutilization is paradoxical given the current challenges facing the food industry, especially the meat sector, which is confronted with the urgent need to replace synthetic additives with "clean label" alternatives, highly demanded by today's consumers [6]. Meat products are highly perishable matrices where the oxidation of lipids and proteins not only deteriorates sensory properties and color but also generates compounds potentially harmful to health [7, 8]. In this context, elderberries emerge as an exceptional solution; while in other industries they are used merely as a colorant, in the meat sector their value lies in their combined action, where their potent antioxidant capacity is even more critical than their color contribution for ensuring and increasing the shelf life of meat products [9, 10]. Therefore, its industrial potential is enormous due to this dual functionality, yet its application outside the scope of beverages or confectionery remains, to date, practically residual or nonexistent.

However, it is important to highlight that the incorporation of natural elderberry extracts presents considerable technical challenges, as their bioactive compounds are extremely sensitive to factors such as light, oxygen, and temperatures during food processing. To overcome these limitations, encapsulation via spray-drying stands out as the most effective and scalable technology [11, 12]. By transforming the liquid extract into a stabilized powder—generally using encapsulating agents such as maltodextrin—the degradation of active molecules is limited [13], while simultaneously facilitating transport, handling, and precise dosing within the food production plant [14, 15]. Furthermore, microencapsulation allows for a controlled and progressive release of antioxidants [15], ensuring that the protective effect remains active for a longer period throughout the product's shelf life and storage.

Despite these advantages, current scientific studies show a high degree of dispersion in the methods of obtaining and presenting elderberries, leading to results that are difficult to compare and hindering industrial standardization. Many of the extraction techniques currently employed are not optimized, sometimes resorting to extreme conditions that degrade the bioactive compounds. Consequently, this study is proposed as a vital point by developing a simple, reliable, and reproducible protocol for the extraction, stabilization, microencapsulation of bioactive compounds from *Sambucus nigra* berries, and their use in the meat industry. The technological development, the product (encapsulated elderberry extract), and their use in meat industry are protected by a utility model (*see* **Note 1**). The goal is to provide a validated methodology that enables the food industry to efficiently leverage this natural resource, ensuring the creation of healthier, more sustainable products with superior commercial stability.

2 Materials

Prepare all solutions using distilled water and/or analytical grade reagents. Prepare and store all reagents at room temperature (unless indicated otherwise). Diligently follow all waste disposal regulations when disposing of waste materials (*see* **Note 2**). Standard laboratory equipment (glassware, containers, glass and plastic test tubes, spatulas, etc.) was used throughout the process, depending on the specific task.

2.1 Bioactive Compounds Extraction Step

1. Ethanol solution (50%) at pH 2: Dissolve 1 L absolute ethanol in 1 L of distilled water and adjust the pH with 1 M HCl.
2. Freeze-drier Lyovapor L300, Büchi; Barcelona, Spain (primary drying pressure limit: 0.500 mbar, time: 240 min; secondary drying pressure limit: 0.400 mbar, time: 120 min).
3. Orbital shaker (Ovan, Mod. OL30-HE, Barcelona, Spain) equipped with incubator chamber with temperature control (Ovan, Mod. IC10-OE, Barcelona, Spain).
4. Ultraturrax homogenizer (IKA T25 digital, IKA®-Werke GmbH & Co. KG, Staufen, Germany).
5. Rotary evaporator (Büchi heating bath B-490 and Büchi rotawapor R-200; Büchi Ibérica, Barcelona, Spain).
6. Centrifuge (Beckman Coulter mod. Allegra™ X-22R, Madrid, Spain).
7. Analytical balance, with a resolution of 0.0001 g (Sartorius, mod. ME614S, Göttingen, Germany).
8. Precision balance, with a resolution of 0.01 g (Sartorius, mod. TE612, Göttingen, Germany).
9. Qualitative filter paper (Filter-lab 1238, Barcelona, Spain).
10. Rotary grinder (Foss KN295 Knifetec, Barcelona, Spain).

2.2 Microencapsulation Step

1. Maltodextrin (DE 10–20; Ceylan S.L., Valencia, Spain).
2. Spray-dryer (Mini Spray-Dryer B-290; Büchi Ibérica, Barcelona, Spain).
3. Ultraturrax homogenizer (IKA T25 digital, IKA®-Werke GmbH & Co. KG, Staufen, Germany).

2.3 Beef Manufacture

1. Beef meat.
2. Sodium erythorbate.
3. Polystyrene trays (25×18×5 cm).
4. Polyethylene film (thickness 50μm)[oxygen permeability <5 mL/(m^2 bar/day); water vapor transmission rate <5 g/(m^2 bar/day].
5. Heat sealer (LARI3/Pn T-VGR- SKIN, Ca.Ve.Co., Palazzolo sull'Oglio, IT, Italy) (*see* **Note 3**).

6. Refrigerated mincer machine (La Minerva A/E22R, Bologna, Italy).
7. Meat mixer (RM20, Mainca, Barcelona, Spain).
8. Burger forming machine (A2000, Gaser, Girona, Spain).

3 Methods

3.1 Elderberry Preparation

Carry out all procedures at room temperature unless otherwise specified.

1. Collect elderberries from wild trees, ensuring a diverse and representative initial batch.
2. Manually select only the berries at an optimal state of ripeness. Systematically discard any unripe, green, or decayed fruit to ensure sample quality.
3. Wash the selected berries thoroughly with water two consecutive times. Store the cleaned berries under refrigeration for a period of less than 24 h.
4. Lyophilize elderberries using a freeze dryer (*see* **Note 4**) to stabilize the samples for further testing.
5. Finely grind the lyophilized material to achieve a uniform powder consistency.
6. Vacuum-pack the ground samples and store them at room temperature (<1 week of storage). If extraction is not performed within 1 week, the samples may be stored frozen at −20°C for a maximum of 6 months.

3.2 Anthocyanins Extraction Procedure

The solid–liquid extraction process should be conducted following the procedure described by Dominguez et al [16], with minor modifications.

1. Mix the lyophilized elderberry with a 50% ethanol solution (pH adjusted to 2 using HCl) in a 1:20 solid-to-solvent ratio (e.g., 25 g per 500 mL) within an Erlenmeyer flask (*see* **Note 5**). Seal the flask to avoid solvent loss through evaporation.
2. The mixture was placed in an orbital shaker at 150 rpm and a controlled temperature of 60°C for 5 h (*see* **Note 6**).
3. Centrifuge the extracts at 3400 × *g* for 10 min at room temperature to separate insoluble solids like fiber, pulp, and skin. Then, filter the supernatant through filter paper (pore size 20–25 μm) to eliminate any remaining insoluble components (e.g., seeds) from the hydro-alcoholic solution.

4. Concentrate the filtered extract by evaporating the ethanol in a rotary evaporator (vacuum) at 75°C for 25 min.
5. Upon completion, the resulting concentrated extract (~5.4 g total solids/100 mL) may be kept chilled for up to 1 week or stored in a freezer until the encapsulation stage.

3.3 Encapsulation Phase: Spray-Drying Procedure

The next stage involved the encapsulation of the concentrated elderberry extract. This was achieved through spray-drying technology, using food-grade maltodextrin as the encapsulating agent. Figure 1 shows, in a simplified and schematic way, the steps of the elderberry anthocyanin extraction and encapsulation procedures.

1. Dissolve 13.5 g of maltodextrin per 100 mL of aqueous elderberry extract [1:2.5 (w/w) core-to-encapsulant ratio]. The mixture was homogenized using an UltraTurrax at 10,000 rpm for 30 min.

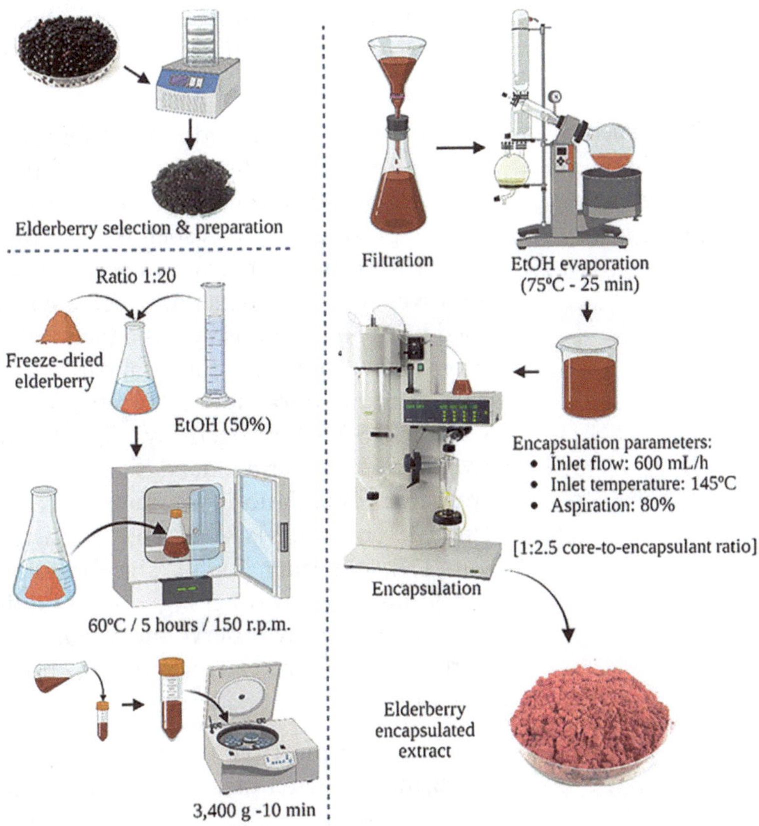

Fig. 1 Schematic and simplified representation of the steps of the elderberry anthocyanin extraction and encapsulation procedures

2. The solution was processed in a spray dryer using an inlet flow of 600 mL/h, an inlet temperature of 145°C, and 80% aspiration (*see* **Note 7**).
3. Recover the encapsulated powder from the cyclone, vacuum-pack and store at room temperature (<1 month) (store in a cool, dry place, protected from direct light) or under freezing conditions (−20°C, <1 year) (*see* **Note 8**).
4. The final encapsulated extract presented a pinkish tone that shifts to an intense red when dissolved. This property renders it an ideal natural dye for the meat industry, complemented by its high antioxidant activity. This antioxidant capacity was evaluated using five different methods, confirming its high antioxidant power. The results showed a DPPH value of 19,013 μg Trolox/g, ABTS of 4998 mg ascorbic acid/100 g, FRAP of 38,217 μmol Fe^{2+}/100 g, ORAC of 109 mg Trolox/g, and an IC_{50} (DPPH) of 11.9 mg/mL. This activity is primarily attributed to the high concentration of phenolic compounds (notably rutin, quercetin, trans-5-O-caffeoylquinic acid, and cis-5-O-caffeoylquinic acid) and anthocyanins (cyanidin-3-O-glucoside, cyanidin-3-O-sambubioside, cyanidin-3-O-sambubioside-5-O-glucoside, and cyanidin-3-O-sambubioside-O-hexoside) [3, 16].

3.4 Preparation of Beef Burgers with Encapsulated Elderberry Extract

The suitability of the food additive for use as a preservative (colorant and antioxidant) in the production of meat products was evaluated by its inclusion in the formulation of beef burgers (a highly perishable product), and its shelf life was monitored for a total of 13 days. Four different batches were prepared: a control batch without additives (Con -), a batch to which a synthetic antioxidant additive used by the meat industry was added (500 ppm of sodium erythorbate) (ERI), and two additional batches to which the encapsulated extract additive was added (at rates of 2.5 g/kg (SN2500) and 5 g/kg (SN5000), respectively). To obtain statistically representative data, the experimental design was replicated three times using the same formulation but starting from different raw materials (in different weeks). All batches consisted solely of beef (91.5%), salt (1%), and water (7.5%). This formulation was selected without the addition of any other ingredients (additives, seasonings, spices, etc.) to exclusively observe the antioxidant and coloring potential of the encapsulated extract additive, ensuring no interference from other ingredients or parameters.

1. The beef meat (containing ~5% fat) was ground through an 8-mm diameter plate under refrigerated conditions (2°C).
2. Salt (1%) was added to the mixture. For the antioxidant-treated batches, each additive (synthetic or encapsulated elderberry extract) was previously diluted in water and then incorporated

into the formulation. Each mass was mixed thoroughly until a homogeneous meat matrix was obtained.

3. Burgers were formed using a burger forming machine (10 cm diameter, 1 cm height; ~100g).
4. The beef burgers were then packaged in plastic trays under modified atmosphere conditions (*see* **Note 3**) and stored under refrigeration (2–4°C) under lighting conditions that simulated supermarket display shelves.
5. Sampling and analysis of the burgers were conducted on the day after packaging (day 1) and after 5, 9, and 13 days of storage.

The chemical composition results reflect that there are no significant differences in moisture, fat, protein, and ash content, or in the fatty acid profile of the burgers across any of the formulations (Table 1) (*see* **Note 9**). This result is both significant and desirable, as it confirms that the use of the elderberry encapsulated extract did not produce any modifications in the chemical composition of the product. Consequently, its application does not alter the nutritional quality of the meat product in any way.

The beef burgers are characterized by a high protein content (~20%) and low-fat content (~5.3%), consistent with the raw material used. The predominant fatty acid was oleic acid (C18:1n-9), followed by palmitic (C16:0) and linoleic acid (C18:2n-6), a profile extensively documented in beef meat and fat.

The inclusion of the encapsulated extract produced a significant protective effect against oxidative processes. In this regard, the lipid oxidation values (measured as malonaldehyde content; Thiobarbituric Acid Reactive Substances or TBARS) and protein oxidation (measured as carbonyls) showed highly promising results (*see* **Note 10**).

As observed in Fig. 2a, the initial lipid oxidation values (day 1) were similar among the batches formulated with an additive (both the synthetic one and encapsulated elderberry extract), while the control burgers exhibited higher oxidation. This demonstrates that the protective power of these additives is already significant during processing and not only during storage.

Regarding the evolution of these values, the extraordinary antioxidant effect of the encapsulated extract must be highlighted. Throughout the remaining days of the shelf life study, the lipid oxidation values were consistently much lower than the control samples and—more importantly—lower than the samples formulated with the synthetic antioxidant additive. In fact, on days 5, 9, and 13, the lipid oxidation values for the samples containing the encapsulated extract (at both 2.5 and 5 g/kg) were nearly half those of the other two batches.

The threshold at which consumers begin to perceive rancid odors and flavors is generally accepted to be between 2 and 2.5 mg

Table 1
Chemical composition and fatty acid profile of beef burgers

	Batch				
	Con -	*SN2500*	*SN5000*	*ERI*	*p Value*
Chemical composition (g/100 g)					
Moisture	71.76±1.24	72.45±0.66	72.25±0.59	72.61±0.61	0.0759
Fat	5.64±0.82	5.19±0.64	5.36±0.68	5.25±0.45	0.3638
Protein	20.37±0.67	20.03±0.64	19.82±0.44	19.95±0.56	0.1378
Ash	1.98±0.06	1.94±0.03	1.95±0.05	1.96±0.06	0.1677
Fatty acids (g/100 g of total fatty acids)					
C14:0	1.44±0.10	1.49±0.11	1.50±0.08	1.46±0.11	0.5033
C14:1n-5	0.15±0.09	0.18±0.08	0.17±0.08	0.16±0.08	0.8568
C15:0	0.16±0.04	0.17±0.03	0.17±0.03	0.16±0.03	0.7524
C15:1n-5	0.44±0.09	0.46±0.11	0.45±0.07	0.45±0.07	0.9750
C16:0	23.08±0.47	23.04±0.51	23.11±0.49	23.07±0.49	0.9880
C16:1n-7	2.26±0.21	2.35±0.19	2.33±0.18	2.28±0.20	0.6187
C17:0	0.54±0.14	0.56±0.13	0.57±0.13	0.55±0.12	0.9625
C18:0	10.89±0.48	10.93±0.51	10.93±0.49	10.90±0.50	0.9956
9t-C18:1	0.16±0.02	0.17±0.03	0.16±0.02	0.15±0.04	0.8298
11t-C18:1	0.84±0.57	0.96±0.61	0.97±0.60	0.84±0.52	0.9113
C18:1n-9	38.04±0.82	38.12±0.87	37.99±0.84	38.04±0.74	0.9843
C18:1n-7	2.45±0.03	2.46±0.05	2.44±0.04	2.44±0.04	0.7412
C18:2n-6	15.42±1.26	15.04±1.31	15.13±1.24	15.35±1.30	0.8644
C18:3n-3	0.80±0.09	0.78±0.09	0.79±0.08	0.79±0.08	0.9671
9c,11t-C18:2	0.12±0.01	0.13±0.02	0.13±0.02	0.12±0.02	0.1589
C20:0	0.16±0.03	0.16±0.02	0.16±0.02	0.16±0.02	0.9943
C20:1n-9	0.68±0.06	0.67±0.07	0.67±0.06	0.68±0.07	0.9036
C20:2n-6	0.57±0.05	0.56±0.05	0.56±0.05	0.57±0.05	0.7730
C20:3n-6	0.23±0.03	0.24±0.03	0.24±0.04	0.24±0.04	0.9520
C20:4n-6	0.73±0.07	0.75±0.07	0.76±0.08	0.77±0.10	0.6200
C20:3n-3	0.11±0.01	0.11±0.01	0.11±0.01	0.11±0.01	0.6057
C20:5n-3	0.12±0.08	0.12±0.08	0.12±0.07	0.12±0.07	0.9988
C22:5n-3	0.25±0.10	0.24±0.10	0.23±0.08	0.24±0.08	0.9710
SFA	36.48±0.84	36.54±0.97	36.63±0.86	36.50±0.92	0.9769
MUFA	45.06±1.57	45.38±1.59	45.21±1.54	45.08±1.40	0.9512
PUFA	18.46±1.46	18.08±1.52	18.17±1.42	18.43±1.43	0.8922
n-3	1.33±0.27	1.30±0.27	1.30±0.25	1.31±0.24	0.9945
n-6	17.14±1.24	16.78±1.32	16.87±1.23	17.12±1.28	0.8641
n-6/n-3	13.32±2.14	13.29±2.18	13.31±1.98	13.41±2.04	0.9991

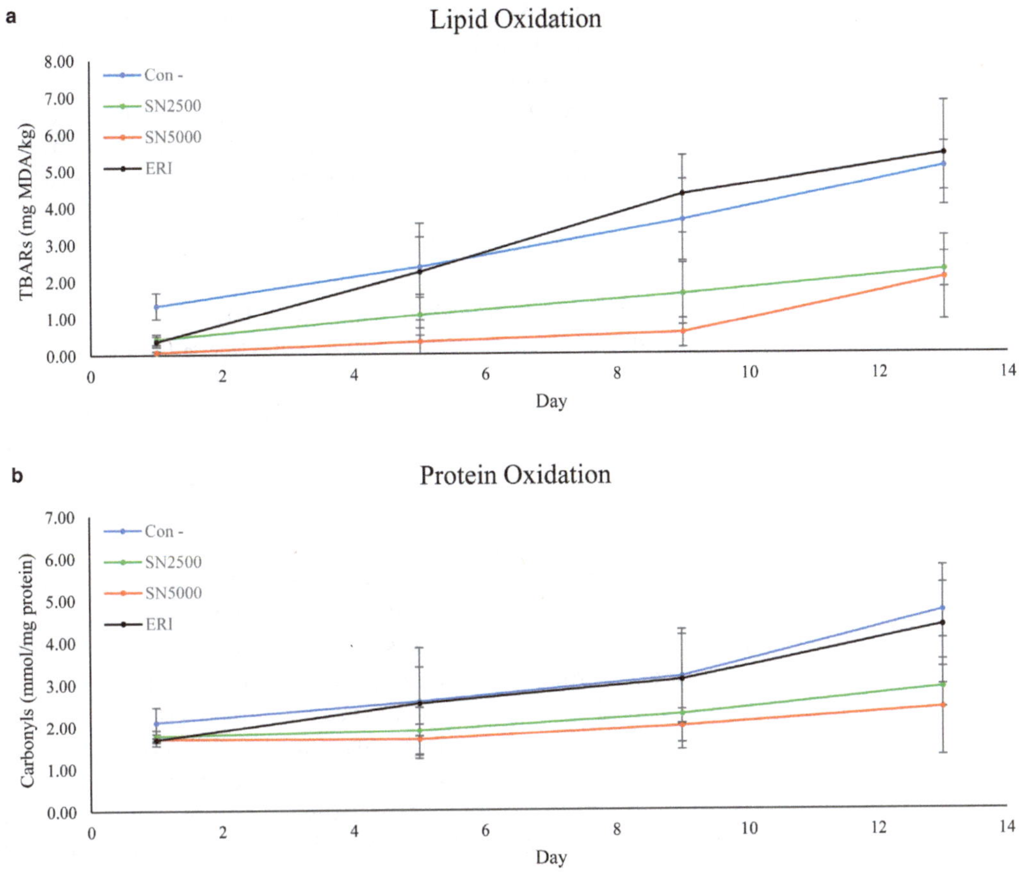

Fig. 2 Evolution of lipid oxidation (TBARS) (**a**) and protein oxidation (carbonyls) (**b**) in beef burgers during the shelf-life study

malonaldehyde/kg. In this study, this threshold was reached by both the control and synthetic antioxidant batches by day 5; therefore, degradation is expected to be perceptible to consumers from this day forward. In contrast, the batches containing the SN2500 (2.5 g/kg) and SN5000 (5 g/kg) did not reach these values until the end of the shelf-life study (day 13). This represents an 8-day increase in shelf life compared to the control and synthetic antioxidant batches.

In a nearly identical manner, the protein oxidation results (Fig. 2b) show a clear protective effect of the additives during processing (day 1). From day 5 onwards, the increase in protein oxidation was much higher in the control and synthetic additive batches compared to the burgers formulated with the encapsulated extract additive.

Considering these results, and in line with the antioxidant capacity data of the additive, it can be affirmed that the additive comprising an encapsulated elderberry extract—prepared under the conditions previously described—possesses a massive protective and antioxidant effect. Therefore, it can be included in meat product formulations to limit oxidative processes in a dose-dependent manner. Furthermore, a superior antioxidant power to that of the synthetic antioxidant was observed, positioning it as a potential substitute for synthetic antioxidants used in the meat industry. This is demonstrated, for instance, by the fact that the degradation processes and oxidation values (lipid and protein) seen in the control and synthetic samples on day 5 were not reached by the samples containing the encapsulated elderberry extract until the very end of the study (day 13).

Another important application of the encapsulated elderberry extract is its coloring power; naturally, this extract presents an intense red coloration that aligns with the desired and expected appearance in meat products and derivatives. It is well known that color plays a vital role in consumer acceptance and, therefore, influences purchase intent and the commercial success of the product. In both fresh meat and meat products, this coloration varies between an intense light red and a dark red, depending on the raw material, and it is linked to myoglobin. This protein (myoglobin), which contains a heme group (iron), presents various shades depending on the oxidation state of the heme group, ranging between red, purple, or brown. Upon oxidation, oxymyoglobin (red) is converted into metmyoglobin (brown), resulting in brown tones and discolorations that lead to direct consumer rejection and, consequently, limit the shelf life of the product. To combat these discolorations, it is common to use artificial/synthetic colorants which, even when metmyoglobin is present, mask the discolorations and provide the product with an attractive color.

Color studies of the burgers were performed using a colorimeter (*see* **Note 11**). Both the total color variation (represented as ΔE) and the changes of the burger color (represented as L^*, a^*, and b^*) were observed, yielding a clear positive effect from the encapsulated extract (Fig. 3).

The parameters of luminosity (L^*) and yellowness (b^*) exhibited a decrease following the addition of the encapsulated extract, likely due to the inherent color and pigments of the elderberry. This trend remained consistent across all burger batches, suggesting a uniform physical impact on the meat matrix. In contrast, the redness index (a^*) showed a distinct behavior between the control group and those containing antioxidants. Thus, the a^* parameter serves as a functional indicator of the extract's ability to inhibit myoglobin oxidation and preserve the characteristic fresh color of the product. Very similar values are observed between the batches formulated with antioxidant additives (synthetic or the encapsulated extracts), demonstrating once again the protective effect of the encapsulated extract in increasing the color stability of the product.

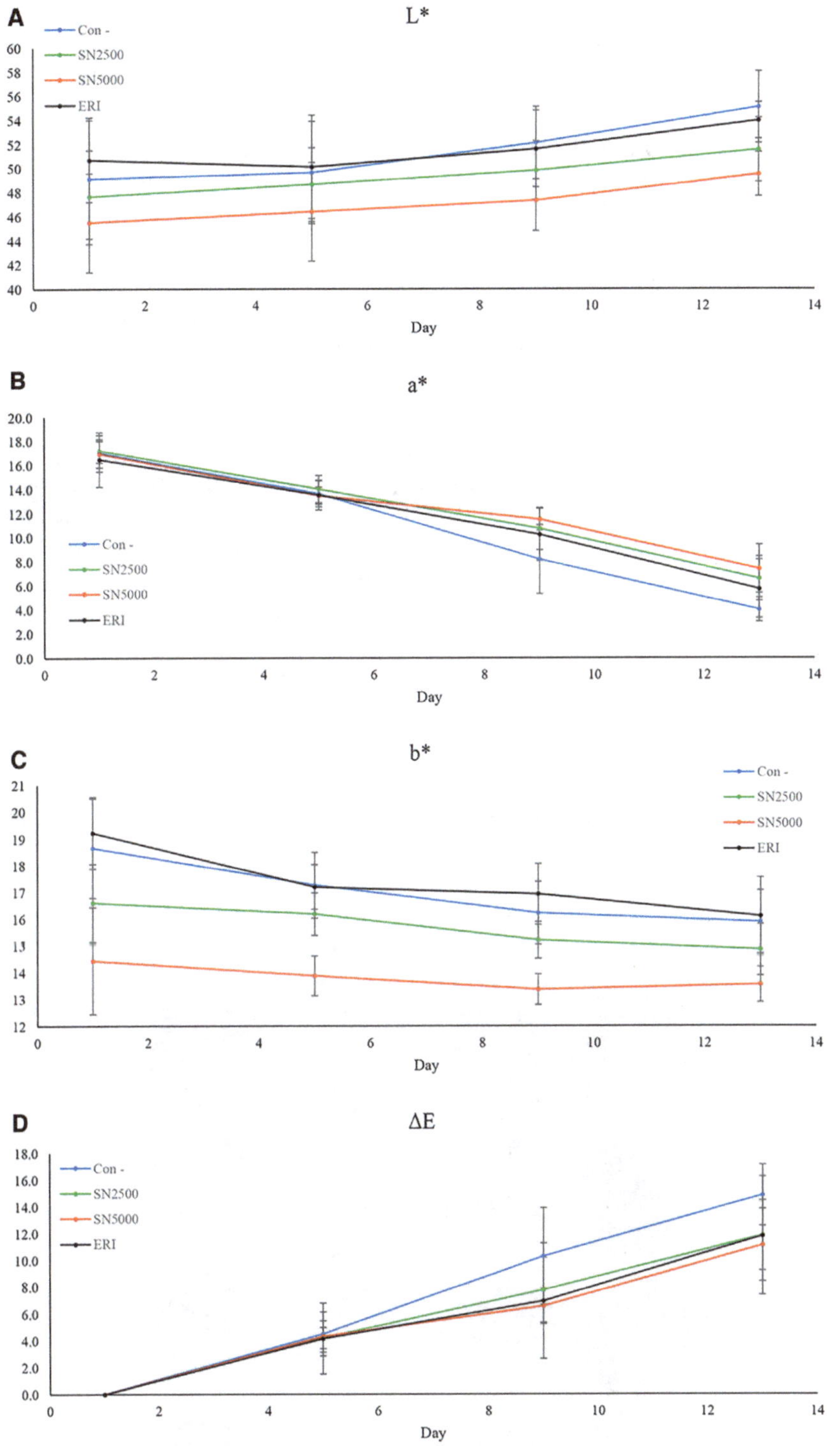

Fig. 3 Evolution of color parameters [(Luminosity **(a)**; Redness index **(b)**; Yellowness index **(c)**; Color variation **(d)**] in beef burgers during the shelf-life study

Similarly, color variation objectively represents, in a single value, how much the redness index, yellowness index, and lightness of the different batches vary throughout the shelf life relative to day 1. In this case, the control batch presented a greater variation than the batches formulated with an antioxidant additive, demonstrating that color is stabilized through their use. Furthermore, the values for the batches using the encapsulated extract and those using the synthetic additive showed no significant differences; therefore, it can be concluded that the natural possesses a potential similar to that of the synthetic additive regarding the protection of the meat product's color. The color-stabilizing activity (both for the red index and total color variation) of the extract is linked, on one hand, to its own red pigmentation, and also closely linked to the antioxidant effect of the extract, which, as previously seen, reduces oxidative processes; this encourages the myoglobin not to oxidize into metmyoglobin and cause discolorations. This fact results in both the color and general appearance of the burger reformulated with the elderberry encapsulated extract being better and more stable for a longer period, which translates into an increase in the product's shelf life. During the shelf life study, an evaluation of the surface discoloration of the burgers was also conducted. In this case, it was observed that the batches reformulated with antioxidants (synthetic or extract)—and more specifically the batch reformulated with 5 g/kg of encapsulated extract—presented less surface discoloration, mainly on days 9 and 13 (Fig. 4). This is in perfect alignment with the color variation and red color intensity measured instrumentally and discussed previously.

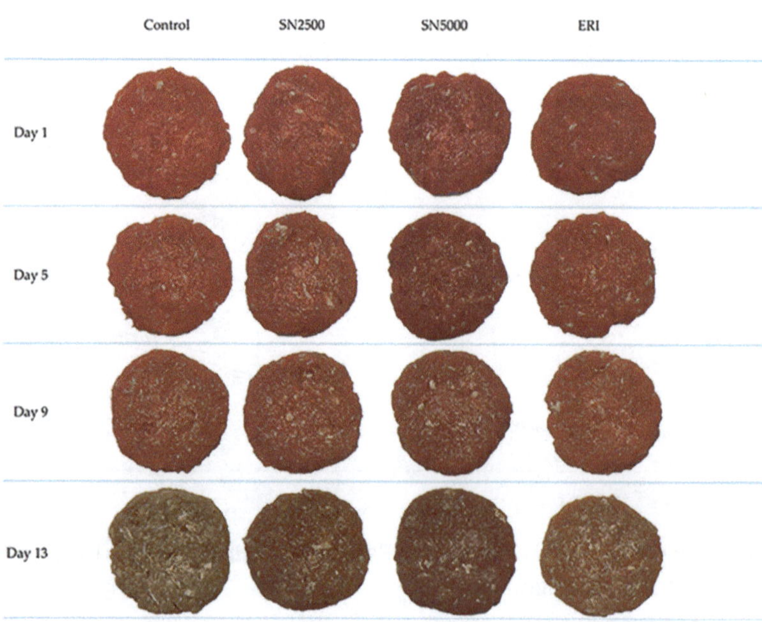

Fig. 4 Evolution of the visual appearance of beef burgers during the shelf life study

This demonstrates that the protective power against discoloration is perceived and positively valued by the consumer, which results in an increase in the shelf life of the product.

Regarding the microbiological stability of the beef burgers, the analysis of aerobic mesophilic bacteria, lactic acid bacteria, and *Pseudomonas* showed a progressive and significant increase over the storage period (*see* **Note 12**); however, no significant differences were observed between the control and treated batches for any of the microorganisms studied (Fig. 5). Aerobic mesophiles rose from initial values of approximately 4.5 log CFU/g to ~6.3 log CFU/g by day 13, with most lots reaching the legal limit of 6.0 log CFU/g only at the end of the shelf life, except for the ERI batch at day 9. Lactic acid bacteria followed a continuous growth pattern until day 9 before plateauing, while *Pseudomonas* populations exhibited a less pronounced increase. This behavior is to be expected since the formulation, as previously indicated, does not contain any of the antimicrobial additives typically used in this product (e.g., nitrites) to prevent interference in the evaluation of the extract's antioxidant and coloring capacity. Consequently, it can be concluded that the encapsulated extract is microbiologically inert under these conditions, as it neither possesses direct antimicrobial properties nor promotes microbial proliferation, thereby maintaining the product's baseline safety profile without interference.

On the other hand, on day 1, the acceptability of visual appearance, cooked odor, texture, flavor, and overall acceptability attributes was evaluated for each sample (Table 2). An acceptance test was conducted using a 7-point hedonic scale, where participants selected the description that best reflected their opinion of each sensory attribute and their overall acceptability. Additionally, preference for each attribute and the overall rating of the beef burgers was assessed through a ranking test (UNE-ISO 8587:2010), assigning scores from 1 (least preferred) to 4 (most preferred). To determine the existence of significant differences between batches, the Friedman Test was applied to the ranking data based on the number of samples and participants. The results presented in Table 2 demonstrate that none of the studied attributes showed significant differences between batches. Similarly, the Friedman Test revealed no significant differences in preference for any attribute. Consequently, it is important to highlight that the incorporation of the encapsulated extract did not exert any effect—positive or negative—on the sensory quality (acceptability or preference) of the burgers. This indicates that the extract can be added without causing organoleptic variations, a vital factor in any food reformulation process.

In summary, it can be affirmed that the encapsulated elderberry extract exerts an extraordinary antioxidant effect, protecting meat products against both lipid and protein oxidative processes more effectively than conventional synthetic additives. This antioxidant

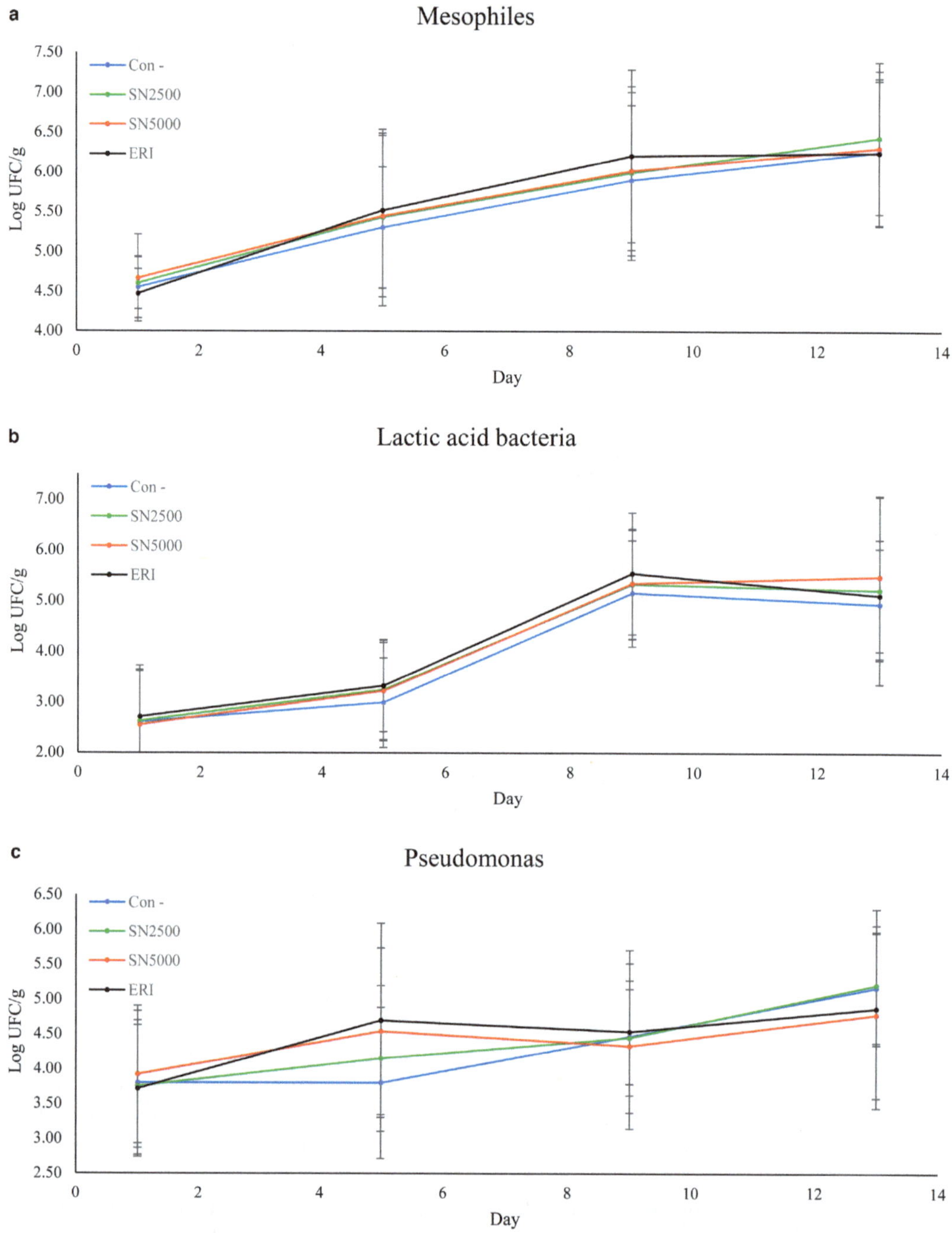

Fig. 5 Evolution of microbial in beef burgers during the shelf life study. Mesophiles (**a**), lactic acid bacteria (**b**), and pseudomonas (**c**)

effect is not only vital throughout the product's shelf life; a clear protective impact was also observed during the processing and

Table 2
Acceptability of the studied attributes in beef burgers

	Batch			
	Con -	*SN2500*	*SN5000*	*ERI*
Visual aspect	5.1	5.6	5.0	6.1
Cooked odor	5.4	5.4	5.3	5.3
Texture	5.6	5.9	5.0	5.6
Flavor	5.6	5.5	5.3	5.3
Overall acceptability	5.5	5.5	5.3	5.3

formulation of the meat product—a critical factor in preventing rapid initial deterioration.

Similarly, the natural additive maintains the desirable color of meat products and derivatives for a longer duration. This is achieved through both its antioxidant capacity and the unique natural pigmentation of the encapsulated extract, which promotes prolonged color stabilization, particularly during storage.

Interestingly, the inclusion of the encapsulated elderberry extract did not produce any nutritional modifications, presenting the same chemical composition and fatty acid profile across all burger batches. It also showed no effect on microorganism development during the shelf life study. Crucially, the reformulation did not influence any of the studied sensory attributes, nor acceptability or preference, which is a critical aspect when bringing a consumer-demanded product to market.

Therefore, based on the obtained results, it can be concluded that this natural food additive possesses high potential as a partial or total substitute for certain synthetic additives currently used in the meat industry, while simultaneously extending the shelf life of meat products.

4 Notes

1. The technological development and the encapsulated elderberry extract presented in the following sections are protected (ES 1 300 302 U) by a utility model (Application number: U202232132/publication date: 6 June 2023). The protection includes the process, the final product obtained (encapsulated elderberry extract), and its use as an additive in the meat industry. Therefore, the steps described below

are included in the invention protection but do not limit its scope.

2. Due to the toxic and/or corrosive nature of the solvents and reagents used in the extraction of elderberry compounds, it is necessary for the operator to take the necessary protective measures (gloves, glasses, etc.), as well as to carry out all the operations in laboratory fume hoods.
3. Burgers packed in MAP conditions (80:20 O_2:CO_2)
4. Lyophilization at −30°C and 0.200 mbar until constant pressure was reached (approximately 2 days).
5. To ensure suitability for food applications, food-grade solvents were utilized for both extraction and encapsulation processes.
6. Due to the presence of toxic cyanogenic glycosides in elderberries, thermal treatment at 60°C for 5 h is required to ensure their complete decomposition. Additionally, the initial selection of fully mature berries is a vital safety measure, as they are naturally free of such compounds.
7. The spray dryer must be preconditioned before use. To do this, turn on the equipment and heat the spray chamber to the desired temperature. Once it reaches 110°C, introduce distilled water to equilibrate the system and stabilize the temperatures. Maintain the final operating conditions with water until the equipment remains steady, at which point the water can be replaced with the extract for encapsulation.
8. To determine the process yield, the final weight of the encapsulated extract is measured against the initial total solids (extract solids plus maltodextrin). The yield is calculated as a percentage, with a minimum requirement of >75%.
9. The chemical composition was evaluated following the procedure described by Bermúdez et al [17], while the fatty acid profile was determined via gas chromatography according to the protocol established by Domínguez et al [18].
10. The determination of lipid oxidation and protein oxidation was carried out following the procedures described by Vyncke [19] and Mercier et al [20], respectively.
11. Color parameters were measured in the CIELAB color space using a portable CM-600d colorimeter (Konica Minolta Sensing Inc., Osaka, Japan). The measurements were taken with a 10o viewing angle, an 8 mm aperture size, and a pulsed xenon arc lamp filtered to illuminant D65 lighting. The colorimeter was calibrated using a white ceramic tile according to the manufacturer's recommendations.

12. Microbiological determinations were performed according to the procedures described by Moraga-Babiano et al [21].

Acknowledgments

Noemí Echegaray and Rubén Agregán acknowledge GAIN for their postdoctoral grants (IN606C-2025/001 and IN606C-2025/002, respectively). Rubén Agregán, Noemi Echegaray, Rubén Domínguez-Valencia, Mirian Pateiro, and José M. Lorenzo are members of the Competitive Reference Group "FunFood" (Functional Food), funded by GAIN (grant number IN607A2023/01). Aurora Cittadini acknowledges the support of a "Juan de la Cierva" postdoctoral fellowship (JDC2023-051424-I) funded by the Spanish "Ministero de Ciencia, Innovación y Universidades" and the "Agencia Estatal de Investigación" MCIU/AEI/10.13039/501100011033 and the FSE+.

References

1. Costa CP, Patinha S, Rudnitskaya A et al (2022) Sustainable valorization of *Sambucus nigra* L. Berries: from crop biodiversity to nutritional value of juice and pomace 11:104
2. Uhl KR, Fyhrie KJ, Brodt SB et al (2022) Blue elderberry (*Sambucus nigra* ssp. cerulea): a robust and underutilized fruit for value-added products. ACS Food Sci Technol 2:347–358
3. Domínguez-Valencia R, Cittadini A, Pateiro M et al (2023) Elderberry lipophilic and hydrophilic bioactive compounds: characterization and extract encapsulation. Foods 12:4233.
4. Senica M, Stampar F, Veberic R et al (2016) Processed elderberry (*Sambucus nigra* L.) products: a beneficial or harmful food alternative?. LWT - Food Sci Technol 72:182–188
5. Ferreira SS, Silva P, Silva AM et al (2020) Effect of harvesting year and elderberry cultivar on the chemical composition and potential bioactivity: a three-year study. Food Chem 302:125366
6. Moraga-Babiano L, Lucas-González R, Domínguez-Valencia R et al (2025) Encapsulated purple sweet potato peel extract as antioxidant and sustainable colourant to preserve the quality of beef burgers during the shelf life. Food Chem 487:144657
7. Domínguez R, Pateiro M, Munekata PES et al (2022) Protein oxidation in muscle foods: a comprehensive review. Antioxidants 11:60.
8. Domínguez R, Pateiro M, Gagaoua M et al (2019) A comprehensive review on lipid oxidation in meat and meat products. Antioxidants 8:429.
9. Marisa Ribeiro A, Estevinho BN, Rocha F (2019) Microencapsulation of polyphenols—the specific case of the microencapsulation of *Sambucus nigra* L. extracts—a review. Trends Food Sci Technol 1–14
10. Rocchetti G, Becchi PP, Lucini L et al (2022) Elderberry (Sambucus nigra L.) encapsulated extracts as meat extenders against lipid and protein oxidation during the shelf-life of beef burgers. Antioxidants 11:2130.
11. Gagneten M, Corfield R, Mattson MG et al (2019) Spray-dried powders from berries extracts obtained upon several processing steps to improve the bioactive components content. Powder Technol 342:1008–1015
12. Ribeiro AM, Estevinho BN, Rocha F (2019) Spray drying encapsulation of elderberry extract and evaluating the release and stability of phenolic compounds in encapsulated powders. Food Bioprocess Technol 12:1381–1394
13. Mattson MLG, Corfield R, Bajda L et al (2021) Potential bioactive ingredient from elderberry fruit: process optimization for a maximum phenolic recovery, physicochemical characterization, and bioaccesibility. J Berry Res 11:51–68
14. Anusha Siddiqui S, Redha AA, Esmaeili Y et al (2023) Novel insights on extraction and encapsulation techniques of elderberry bioactive compounds. Crit Rev Food Sci Nutr 63: 5937–5952
15. Mehta N, Kumar P, Verma AK et al (2022) Microencapsulation as a noble technique for the application of bioactive compounds in the food industry: a comprehensive review. Appl Sci 12:1424

16. Domínguez R, Zhang L, Rocchetti G et al (2020) Elderberry (*Sambucus nigra* L.) as potential source of antioxidants. Characterization, optimization of extraction parameters and bioactive properties. Food Chem 330:127266
17. Bermúdez R, Echegaray N, Fraqueza MJ et al (2022) Chemical composition. In: Lorenzo JM, Domínguez R, Pateiro M et al (eds) Methods to assess the quality of meat products. Springer Nature, New York, USA, pp 1–16
18. Domínguez R, Pateiro M, Campagnol PCB et al (2022) Fatty acids. In: Lorenzo JM, Domínguez R, Pateiro M et al (eds) Methods to assess the quality of meat products. Springer Nature, New York, USA, pp 41–52
19. Vyncke W (1975) Evaluation of the direct thiobarbituric acid extraction method for determining oxidative rancidity in mackerel. Fette Seifen Anstrichm 77:239–240
20. Mercier Y, Gatellier P, Renerre M (2004) Lipid and protein oxidation in vitro, and antioxidant potential in meat from Charolais cows finished on pasture or mixed diet. Meat Sci 66:467–473
21. Moraga-Babiano L, Lucas-González R, Domínguez-Valencia R et al (2026) Encapsulated prickly pear peel co-product extract: an antioxidant and colourant additive to preserve the quality of beef burgers packaged in modified atmosphere. Meat Sci 233:110020

Chapter 12

Healthy Dry-Cured Sausages: Salchichón as a Model System for Dietary Fiber Application

José Ángel Pérez-Álvarez, Raquel Lucas-González, Alonso Pérez-Fernández, Manuel Viuda-Martos, José Manuel Lorenzo Rodriguez and Juana Fernández-López

Abstract

The technological, microbiological, and techno-functional strategies for dietary fiber addition into fermented dry-cured meat sausages using the Salchichón as a model system are evaluated. Dietary fiber obtained from agro-industrial coproducts represents a promising strategy for improving the nutritional value and functional properties of these types of meat products.

Keywords Salchichón, Fermented dry-cured sausages, Prototype, Dietary fiber, Coproducts valorization, Intermediate Food Products, Microbiota

1 Introduction

One of the most important opportunities for the meat industry is related to the development of healthy meat products by reformulation. This challenge also involves the more traditional sectors of this important agri-food industry, including **fermented dry-cured sausages**, among which *salchichón* stands out. Consequently, any reformulation strategy must carefully balance nutritional improvements with the preservation of the traditional sensory characteristics that define these products. Given the special characteristics of this meat product specialty, reformulation should involve a holistic understanding of the product, considering not only its technological properties but also its anthropological, culinary, gastronomic, sensory, cultural, and social dimensions. This integrative approach allows researchers, technicians, and processors to understand this complex and, well appreciate meat product speciality.

Dietary fiber (DF) is widely recognized as a beneficial dietary component associated with improved digestive health and metabolic

José Manuel Lorenzo Rodriguez (ed.), *Meat Product Quality Improvement*, Methods and Protocols in Food Science,
https://doi.org/10.1007/978-1-0716-5460-6_12,

regulation. However, their perception of fiber in meat products may still be influenced by a complete lack of information about their role in them. The perception of DF gradually changed. Nevertheless, its perception in meat products can still be influenced by the terminology used to describe its origin. In this regard, it is important to distinguish between **waste**, **by-products**, and **coproducts**. From a consumer perspective, the term "waste" is often associated with garbage, while "by-products" may be perceived as low-quality materials mainly used for animal feed. In contrast, coproducts generated during agro-industrial processing can be valorized through several technological processes to produce high-value ingredient, **Intermediate Food Product(IFP)**, suitable for human consumption. DF or IFP can be obtained from vegetal (fruit juices processing; cereal milling; vegetable processing) or animal (crustacean shells, insect exoskeleton) agro-industries (fruit and vegetables) coproducts. In several countries, animal DF addition to meat products is not permitted in meat products.

Dietary fiber is a highly heterogeneous group of compounds that differ not only in their chemical composition (*see* **Note 1**) but also in their physical (*see* **Note 2**), physicochemical, technological, and physiological properties. This diversity makes DF a particularly versatile ingredient for food reformulation strategies [1]. Dietary fibers can be classified according to several criteria, including their physical and purity state, origin, solubility, and major bioactive compounds. In Table 1, several commercial DF solubilities are shown. Also, physicochemical properties, techno-functional behavior (Table 2), and physiological effects are different according to DF or IFP. Dietary fibers could affect fat (fat/oil binding capacity), cholesterol (cholesterol absorption capacity), and bile acid absorption (Bile-holding capacity) during digestion [2]. Viscous fibers or

Table 1
Solubility characteristics, composition in bioactive compounds, and raw material from which they are obtained

Commercial dietary fibers	Solubility	Major bioactive compounds	Raw materials
Viscofiber	Soluble	Oat/Barley	Beta-glucan (50%) (IFP)
Novelosa	Insoluble	Resistant starch	Corn hybrids
Liposan Ultra	Soluble	Chitosan	Crustacean exoskeleton
Slendid	Soluble	Pectin	Citrus, apple, date fruit
Inulin	Soluble	Mixture of poly- and oligosaccharides with nearly the same chemical structure	Chicory roots
Liposterine	Insoluble	Polyphenolic compounds (condensed polyphenols with high molecular weight)	Carob

Table 2
Techno-functional values of Water Holding Capacity and Oil Holding Capacity (g water/g and g oil/g of dietary fiber, respectively) from commercial dietary fibers are industrially used in the meat industry

Commercial dietary fibers	WHCg water/g of dietary fiber	Commercial dietary fibers	OHCg oil/g dietary fiber
Viscofiber	21.61	Justifiber	8.89
Citric-FI 100	12.0	Vitacel Bambú	5.35
Pea (Swelite)	11.04	Guisante (Swelite)	4.04
KMC Pif potato	10.33	Quitosan	3.49
Justfiber (Bambu)	8.50	Liposan Ultra	3.22
Liposan ultra	7.86	Fibrim	3.11
Fibrim (Soy bean)	6.77	Viscofiber	2.86
Vitacel bambu	6.47	Citric-FI 100	2.85
Vitacel potato	6.06	B-Glucano Trades	2.79
B-Glucano trades	5.84	Vitacel Potato	2.67
Quitosan	5.56	Slendid	2.62
Fibrex	4.33	Quitosan Trades	2.58
Quitosan trades	3.75	Fibruline (inulin)	2.55
Novelose 330	2.65	Fibrulose	2.37
Slendid	1.42	Fibrex	2.24
HI-Maize 260	1.41	Novelose 330	2.22
Fibrulose	0	HI-Maize 260	1.73
Fibruline (inulin)	0	KMC Pif Potato	1.72

IFP rich on them (oat β-glucan in oat products and psyllium) are better for these purposes [3]. Technological problems in mixing, stuffing, and maturation stages could be occurred (more kneading force, changes in visual appearance, a fatty aspect in surface color "muddy aspect" and lower weight losses, respectively) among sensorial changes inside sausage (acid taste, nonformed cured meat color, fatty taste) when viscous DF are used.

One of the most important properties for its relationship with processing parameters is the product stability (physicochemical, oxidative, and microbial parameters) and economics. Techno-functional properties of dietary fibers are particularly relevant for their application in meat products. These properties include: (1) water-holding capacity (WHC); (2) oil-holding capacity (OHC); (3) swelling capacity (SC); (4) emulsifying capacity; (5) emulsion stability; and (6) gel formation capacity. For salchichón, the main techno-functional properties to be evaluated in fermented dry-cured sausages are **Water Holding Capacity** (WHC). This property is influenced by particle size (finely milled particles are recommended) to avoid a sandy texture [4, 5]. This defect could be obtained from lignin/lignans rich fibers. In Table 2, WHC and OHC of several DFs used in the meat industry are showed.

Salchichón is a special case in fermented dry-cured sausages since it is formulated with the basic ingredients of a meat product: fatty and lean meats (*see* **Note 3**), salt (*see* **Note 4**), curing agents (*see* **Note 5**), pepper (white or black) (*Piper nigrum L.*) (*see* **Note 6**), casings (natural or synthetic) (see **Note 7**), new casing technology applied to dry-cured sausages (*see* **Note 8**), starter cultures (*see* **Note 9**), or the use of DF/IFP with prebiotics characteristics and probiotic strains, to elaborate symbiotic salchichón (*see* **Note 10**).

The lean and fatty meats ratio depends on salchichón type. Salchichón name depends on geographical origin (Toledo, Málaga) casing diameter (Fuet, Longaniza imperial, Longaniza de Pascua), pork meats breed (white or Iberian), and commercial category ("extra," "primera"). In Fig. 1, some of these different types of salchichón can be observed.

The prototype represents the first complete version of a product during the innovation process and allows researchers to evaluate the effects of new ingredients under controlled conditions. In the case of salchichón, the formulation used in experimental trials is intentionally simplified to avoid possible interference from other ingredients and additives that could mask the technological effects of dietary fiber. For this reason, spices with strong sensory contributions, such as cloves, chilli flakes, paprika, annatto, or saffron, are usually avoided. Instead, only the essential ingredients are used (*see* **Notes 1–6**). The main characteristics of salchichón are caused by several physical, chemical, biochemical, and microbiological processes. The casing diameter, spices mixing, meat endogenous enzyme activities, microbial activities, autoxidation processes, and the interaction among odorous compounds give the special characteristic of this type of product and define the differences between salchichón types.

Other processing variables, such as temperature, drying time, fat content, and sausage diameter, are also carefully controlled to ensure that any observed effects can be attributed to the incorporation of dietary fiber or IFP. Commercial DF and IFP or experimental IFP can be used in salchichón prototypes. Figure 2 illustrates examples of experimental Intermediate Food Products (IFP) "fiber-rich flours" obtained from orange and fresh date coproducts valorization.

DF and/or IFP can be used as functional ingredients not only for their content in soluble and/or insoluble dietary fiber but also for their functional properties (technological and physiological). All these properties are interesting for the selection of a fiber-rich extract as an ingredient in a healthy meat product development. Depending on

Fig. 1 Commercial samples of different Spanish salchichón types with several quality characteristics, degree of mincing, use of colorants, product diameter, and packaging formats

these properties, they can be used as fat replacers, volume enhancers, binder, bulking agent and stabilizer, emulsify agent, etc. [2].

Generally, fermented dry-cured sausages, such as salchichón, are structurally complex food systems. Their production involves a series of simultaneous transformations [6]. Thus, when the incorporation

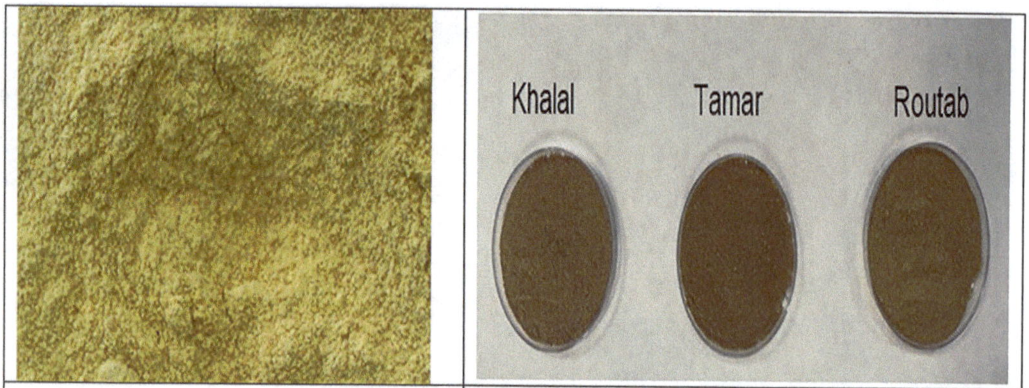

Orange coproducts fibre-rich (60%) flour. Unpublished data	Fresh date coproduct fibre-rich flours (65, 62, 58%) at different maturation stages (Khalal, Routab and Tamar).

Fig. 2 Fiber-rich flours obtained from the valorization of orange juice coproducts and fresh date fruits at different maturation stages (Khalal, Rutab, and Tamar)

of DF/IFP into meat products is studied, it is essential to begin with a model system or prototype. Salchichón is a useful model system for meat product innovation. Its simplicity prototype allows evaluation of the effects of new ingredients under controlled conditions.

2 Materials

This chapter will place special emphasis on those aspects that facilitate the understanding not only of the elaboration process but also those extrinsic and intrinsic aspects of it and those empirical tools or technical/analytical criteria that help technicians/researchers or industrial to properly guide the addition of DF/IFP to improve the nutritional quality and healthiness of these type of meat products that is so demanded and appreciated by consumers, such as Spanish salchichón.

This section will detail according to the salchichón elaboration process, raw materials, changes of various kinds (physical, chemical, technological, microbiological, and/or sensorial), and the effects that both DF and/or IFP may have on the dry-fermented sausage. It will also include the main scientific, technological, industrial, and empirical analyses and recommendations compiled in this chapter. Thus, aspects such as ingredients and raw materials characteristics (*see* **Notes 3–10**); raw materials and dietary fiber empirical, industrial, and analytical measurements (*see* **Notes 11–20**); empirical industrial tips, there are not strictly necessary the use of sophisticated machinery or analytical equipment (*see* **Notes 21 and 22**). Physical, physicochemical changes (*see* **Notes 24–29** and **Notes 30–33**, respectively), and key microbial aspects during the elaboration process (*see* **Note 34**).

One of the most important aspects is the quality of the meat (lean and fatty cuts). The type of meat cuts, their ratio, the percentage of tendons, and collagen-rich cuts affect the final quality of salchichón (*see* **Note 3**). The spices and additives used, as well as the type and diameter of the casing (*see* **Notes 5–7**). These aspects are of great technological and industrial interest, since the interaction between the DF/IFP, the meat matrix, and how this interacts with the environment inside the casing will allow for a product that is at least equal to or better than the traditional one.

3 Methods

Salchichón elaboration process is relatively easy if key processes, such as raw material mixing, fermentation, and dry-curing (also called dry- maturation or ripening) must be considered. The salchichón flow chart can be observed in Fig. 3.

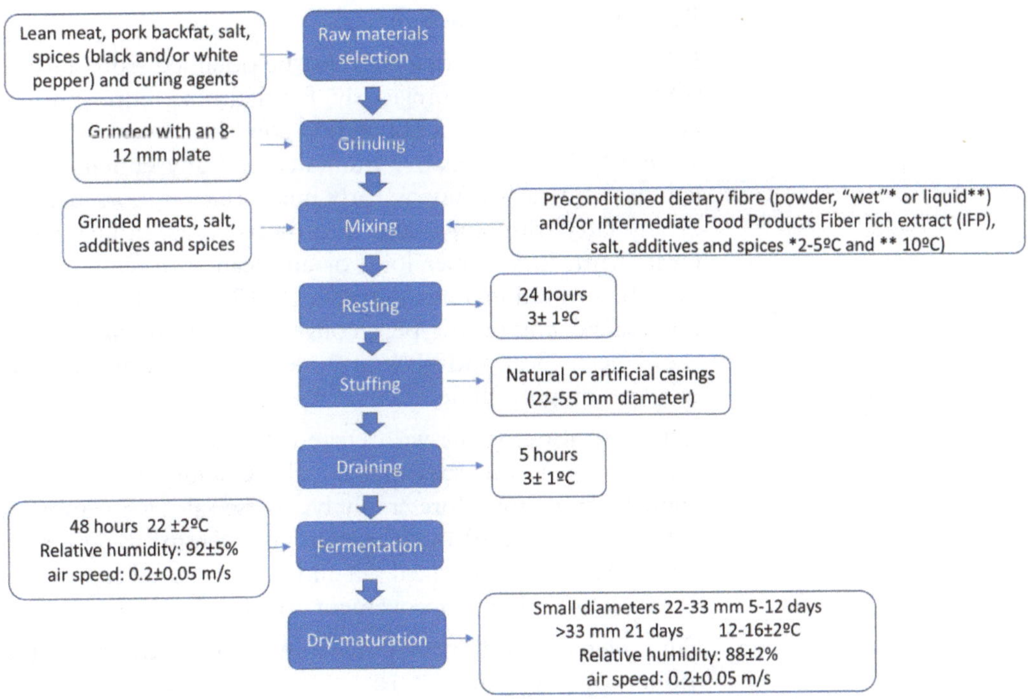

Fig. 3 Dietary fiber added to the prototype Salchichón flow chart

4 Notes

Dietary fibers and Intermediate Food Products

1. Purified dietary fiber (inulin, carboxy-methyl-cellulose-CMC, >98%) or IFP (60%–70%) purity, from coproduct valorization (cereals, apple, date, mango, carrot, rich fiber extract, among others); beta-glucans (50%–70%); solid as powder ("flour"); or liquid (date fiber-rich juice). CMC has an excellent advantage, has cold solubility, but as disadvantage increased thickening (high viscosity). Inulin acts as an effective bioactive compounds carrier.
2. Soluble (low-molecular-weight fructo-oligosaccharides, fiber-rich "date juice"), insoluble (lignan, lignin-rich dietary fibers, tiger nut resistant starch; date pulp), or partially soluble dietary fibers (low, medium, or high methoxyl pectins, etc.) [7, 8]. Most of the DF used in meat products come from pectin-rich IFP. Depending on the functional properties such as gelling ability, water binding ability, and acting as an emulsifier, pectin is one of the natural ingredients for healthy meat product formulations. Pectin interacts with each type of meta protein. Actin, myosin, and collagen interact at the protein chain level. This interaction is with amino acid and H-bonds [9].

   Raw materials characteristics
3. Meat (lean and fatty tissues) must be obtained from authorized slaughterhouse; previously, for their use, both must be kept at 1–4°C. Chamber operation conditions: room temperature <12°C; 95% relative humidity, and any sanitary regulations from each country. Fatty meats (bacon, pork backfat, or any fatty trimmings) and lean meats (pork shoulder lean meat, ham, loin, upper loin, or any lean trimmings). Meat must be ground in a meat grinder (8–12 mm hole diameter). For a salchichón prototype, a conventional meat composition is 60% lean meat and 40% fatty meats. Meat is mixed in an arm kneading machine.
4. Salt must also comply with current legislation, and is overwhelmingly Mediterranean Sea salt. Granulometry is very important; it should preferably be coarse salt (1.5–3 mm) or salting salt (>3 mm) or flaked salt (0.5–5 mm) for gourmet products. The use of flaked salt and demonstrated technological and/or sensory improvement are not found.
5. Curing agents (sodium or potassium nitrites and/or nitrates, sodium or potassium ascorbate or erythorbate) are used for several purposes, including red cured meat color formation (nitroso-myoglobin), *Clostridium botulinum* growth

inhibition, antioxidant properties, and avoid warmed-over flavor. The dosage depends on the legal requirements for each country.

6. Pepper (*Piper nigrum L.*) could be used as entire, coarse, and fine ground (depends on salchichón type: Toledo, Extra, Fuet, etc.) and used inside meat batter or on the salchichón surface. For prototype purposes, fine-grounded pepper is the most suitable. And must be included in the meat batter [10]. Black pepper contains piperine, camphor, carvacrol, eugenol, and ubiquinone as main bioactive compounds, which play an important role in salchichón processing. All act as antioxidant, antimicrobial, and antifungal [11]. Also, it is interesting that black pepper acts as an insect repellent and is used both internal or in the salchichón surface.

7. Casing is part of sausages, giving shape, size, and integrity, but also having a role in volumetric, structural, and chemical changes which occur in sausages during different elaboration steps. Mechanical properties of casing, such as tension strength, elasticity, temperature resistance, transparency, and gloss are responsible for the structural integrity, size, shape, volumetric changes, texture, and appearance of the finished product. The final product is a result of interactions between casing and meat batter filling during all processing steps. Collagen and regenerated cellulose casing facilitate diffusion of oxygen, moisture, and other nutrients such as sugars, amino acids, ions, etc. for the growth of external microbes to produce natural specialty-flavored fermented sausages [12]. Salchichón is dried until reaching the necessary amount of water content (Aw value). The lactic acid generated by lactic acid bacteria (LAB) metabolism reduces the casing permeability. If casing cannot adjust to changes in sausage volume, it can lead to salchichón structural defects. Artificial casings have an advantage, are the better choice, for the production of large diameter sausages, whereas they are equivalent to natural casings, for production of small diameter sausages [13]. Before use casing (especially artificial collagen casing or regenerated cellulose) is essential to follow certain technical recommendations. Thus, proper rehydration is needed. Soak the casings in warm water to facilitate sliding and filling. Time soaking must be controlled by monitoring the immersion time to avoid exceeding the permitted technical limits (see technical specification provided by producer/supplier). The internal lubrication must be done. Run tap (chlorinated) water through the casings inside to improve lubrication and prevent clogging in the sausage stuffer. Correct the stuffing using the already flexible and lubricated casing. Following these recommendations, casing detaching can be avoided.

Also, collagen casing reduces the content of biogenic amines in fermented dry-cured sausages [14].

8. Alginate coatings as a substitute of collagen casing and could be a feasible alternative in fermented dry-cured sausage (salchichón) with small caliber diameters from a technological and safety point of view. Stuffing in alginate coating resulted in faster sausage drying kinetics compared with collagen-casing sausage [15].
9. Lactic acid bacteria (LAB) are recognized for their important technological and sensorial role in dry-cured fermented sausages. There are several commercial salchichón starter cultures, but also another commercial LAB strains—*L. sakei*, *L. plantarum*, and *P. pentosaceus*— also have been demonstrated its potential as probiotic starters for dry-cured fermented sausages showcasing both antibacterial and antioxidant activities. These LAB use of probiotic starters not only maintained the physicochemical attributes and sensory evaluation but also enhanced the products' color and texture [16].
10. Symbiotic dry-cured fermented sausages are a challenge for industrial and research. Experimental approaches make it feasible using prebiotic DF and/or IFP and probiotic bacteria. Probiotic strains maintained high viability and contributed to the suppression of spoilage and pathogenic microorganisms throughout storage [17]. Probiotic fermented sausages not only introduce innovation into the meat industry but also offer consumers a new way to consume probiotics, serving as a promising option for health-conscious individuals [18].

Raw materials and dietary fiber empirical, industrial, and analytical measurements

11. CIELAB color parameters (illuminant D_{65}; standard observer 10°, Guidelines for meat color evaluation from American Meat Science Association [19]) are the best way to measure meat product color. But infinite solid and background are needed to establish [20]. Also, reflectance spectra (360–740 nm) measurement is highly recommended. Dietary fibers with negative a* greenish color (low or high) and positive b* (yellowish color) or negative b* value (blueish color) must be avoided. To make a precheck analysis, a*/b* ratio should be positive (a* and b* positive) with a range value ≈1 or >1. In Fig. 4, reflectance spectra of several salchichón prototypes are added with different IFP (pomegranate, date, and a mixture of both ratios 1:1).
12. pH measurements: The pH value was determined in a suspension resulting from blending a 0.5-g sample with 50 mL deionized water for 2 min using a calibrated pH meter. To make measurements of this parameter, a conventional

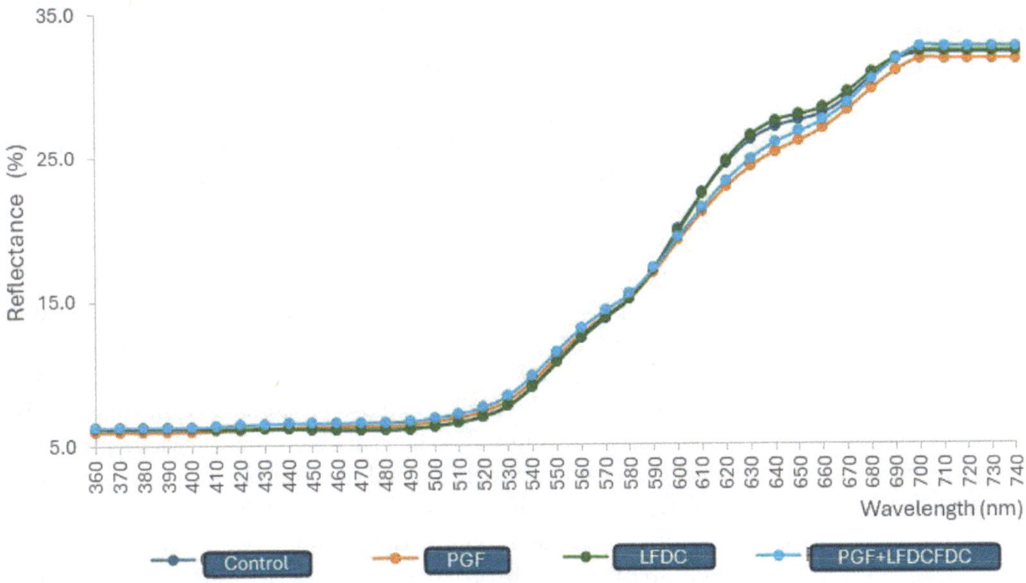

Fig. 4 Reflectance spectra (360-740 nm) in salchichón prototypes (33 mm diameter) formulated with a mixture of pomegranate IFP powder (PGF), liquid date IFP (LFDC), and the mixture of both (PGF+LFDC, ratio 1:1) at the end of the ripening stage (12 days)

electrode or a punction electrode is useful. For sausage pH determination, the punction electrode is more useful. This parameter could be an indirect signal about metabolic microbiota activity.

13. Water activity: This measurement could be made with different devices (Novasina, Decagon, among others). All equipments must be calibrated with supersaturated salts (see equipment's instructions). The most precise and affordable are the Novasina equipments with controlled temperature. Temperature measurements 25°C.
14. Total polyphenol content (TPC): The total phenol content (TPC) could be determined using Folin–Ciocalteu reagent [21]. About 0.3 mL of the ethanol or methanol:acetone solutions of the different samples (30 g/L) are introduced into test tubes, followed by 2.5 mL of Folin–Ciocalteu reagent (diluted 10 times with water) and 2 mL of sodium carbonate (7.5%, w/v). The tubes are vortexed, covered with parafilm, and incubated at 50°C for 5 min. Absorption at 760 nm is measured with a spectrophotometer and compared to a gallic acid calibration curve. The results were expressed as mg gallic acid equivalents (GAE)/g dry mass [2].
15. Antioxidant properties: Thiobarbituric acid (TBA) test extraction method is easier to determine in industry or in small labs. Small adaptations must be done for each meat products. Prior

to sampling, the residual nitrite level must be determined by the standard ISO/DIS 2918 method (ISO, 1975). Check the elaboration conditions to apply sulfanilamide or not. Samples of 4 g (+0.001 g) will be introduced into centrifuge tubes. Then, 10 mL of Tricloroactic acid and phosphoric acid extracting solution (TCA-PA) and, for the MA percent recovery study, 5 mL of MA solution or 5 mL of distilled water were added. All DF or IFP could increase the TBA value (for its processing itself). Tubes were stirred for 5 min with a Heidolph vortex stirrer (model REAX 2000). At the end of this time, 5 mL of TBA solution is added to each tube, which is stirred again for 2 min. Tubes are centrifuged for 15 min, at 3500 rpm, in a centrifuge. To separate the liquid phase. This is filtered through a Whatman No. 1 filter paper into assay tubes. Later, tubes are placed in a thermostatically controlled water bath for 35 min at 100°C, or 15 h in the dark at room temperature, to allow the development of the MA-TBA color complex. The absorbance could be measured in a spectrophotometer at 532 nm. At the same time, a "model system" was processed, as well as in the same way, without salchichón (fiber added or not) prototype. Results are expressed as mg of MA per kg of sample and are calculated following the Witte and co-workers method [22]. In this type of prototype, it is important to check this parameter during the elaboration process; thus, it could prevent that fiber added could act as a prooxidant instead of an antioxidant [23].

16. Dietary fibers could affect technological aspects during processing. Depends on concentration, several types of technical problems in mixing, stuffing, and maturation stages could occur, such as increased kneading force, a "muddy aspect," given by changes in visual appearance, a fatty aspect on surface. Also, color changes and lower weight losses, respectively. Among sensorial changes inside sausage (acid taste, nonformed cured meat color, and fatty taste), when viscous DF are used.

17. Water:fiber ratio: This interaction is very useful and easy to implement in industrial research activities. DF "pure" or IFP (powder) and water (2–8°C), in several ratios (0.5:1, 1:1, 1:2, etc.), are mixed and resting (between 30 min and 24 h under refrigeration conditions). The interaction between dietary fiber (DF) and water influences water absorption and/or adsorption, which plays a key role in determining the consistency and processing behavior of the meat batter and the resulting sausages. These characteristics could potentially become part of the meat matrix once incorporated, and the results show how it will interact with the sausage's meat matrix.

18. Mixing DF or IFP with the prototype salchichón meat batter. This empirical procedure implies that DF or IFP must be added in power and/or water added. Concentration could be indirectly determined with textural and visual appearance on salchichón prototype meat batter.
19. Water Holding Capacity (WHC): The DF samples are measured by the modified centrifugation method. 10 mL of ultrapure water was added to a centrifuge tube containing 300 mg of the DF samples. Subsequently, the centrifuge tubes are stored at 25°C for 18 h. After centrifugation (3000 rpm, 20 min at 25°C), the supernatant is discarded and the pellet is weighed. The WHC of each sample is expressed as the weight of water held by 1 g of the corresponding DF samples [24].
20. Oil Holding Capacity (OHC): The oil-holding capacity (OHC) of DF samples is measured by the modified centrifugation methods described by Robertson and co-workers [24]. 5 g of sunflower oil is added to a centrifuge tube containing 160 mg of the DF samples. Subsequently, the centrifuge tubes are stored at 25°C for 18 h. After the centrifugation (3000 rpm, 20 min at 25°C), the supernatant was discarded, and the pellet was weighed. The OHC of each sample is expressed as the weight of oil held by 1 g of the corresponding DF samples.

    Empirical industrial tips (there are not strictly necessary the use of sophisticated machinery or analytical equipment)
21. Also, it is important to make a precheck mixing DF or IFP with prototype meat batter. Many fibers, after adding water, expose orange-yellow (citrus), greenish (legumes, such as Guandú (*Cajanus cajan*) or Green peas, vegetables such as spinach or celery), orange (carrot fiber), or brown (date fiber), among other colors. Figure 5 shows an example of a preliminary trial incorporating a Guandú (*Cajanus cajan*) and date (*Phoenix dactylifera L.*) paste as IFPs, prior to its potential use as an ingredient to be added in salchichón processing.
22. Effect of DF and/or IFP upon microbial surface. A previous trial could be made using microbial growth on the fermented sausage prototype casing surface. The behavior of DF and IFP could change if it is previously moistened or is added as a powder. Figure 6 an undesirable effect of the carrot fiber (3% and 6%) previously moisturized with a ratio (1:1) during 12 h at 4°C, upon salchichón protype is showed.

    Physical changes during elaboration process
23. Natural casings are typically used in traditional products, whereas artificial casings are used in sliced products because

Fig. 5 Effect upon salchichón meat batter color with the addition of several guandú (*Cajanus cajan L.*), concentrations (0%, 10%, 20%, and 30% and 2% of date paste (*Phoenix dactylifera L.*) IFP as technological feasibility preliminary assays

they show higher water permeability and resistance, constant diameter, and they are easily removed.

24. Shrinkage and casing separating the from the sausage. During stuffing salchichón has the higher amount of water. During the ripening stage (Fig. 6), water removal takes place through water evaporation from the inner to the outer surface. This water loss provokes shrinkage. Dehydration at too high a temperature (>24°C) can increase the risk of casing separation and/or breakage, also an excessive shrinkage. Consequently, a water gradient concentration between the inner part of the solid and its external surface is established [25]. To avoid excessive shrinkage, reformulate with leaner meat (chicken, turkey). Grind the meat coarsely, and dehydrate slowly. Avoid overfilling the casing. Use a high-quality casing (natural or artificial) to prevent tearing. Make shallow cuts (with a needle) in the casing after stuffing if overfilling was produced when salchichón is dehydrated casing separation could occurs.

25. The special salchichón structural complexity, composed of meat and other functional ingredients (DF and/or IFP), reduces thermal conductivity and permeability during the drying process [26]. Therefore, the elaboration process could

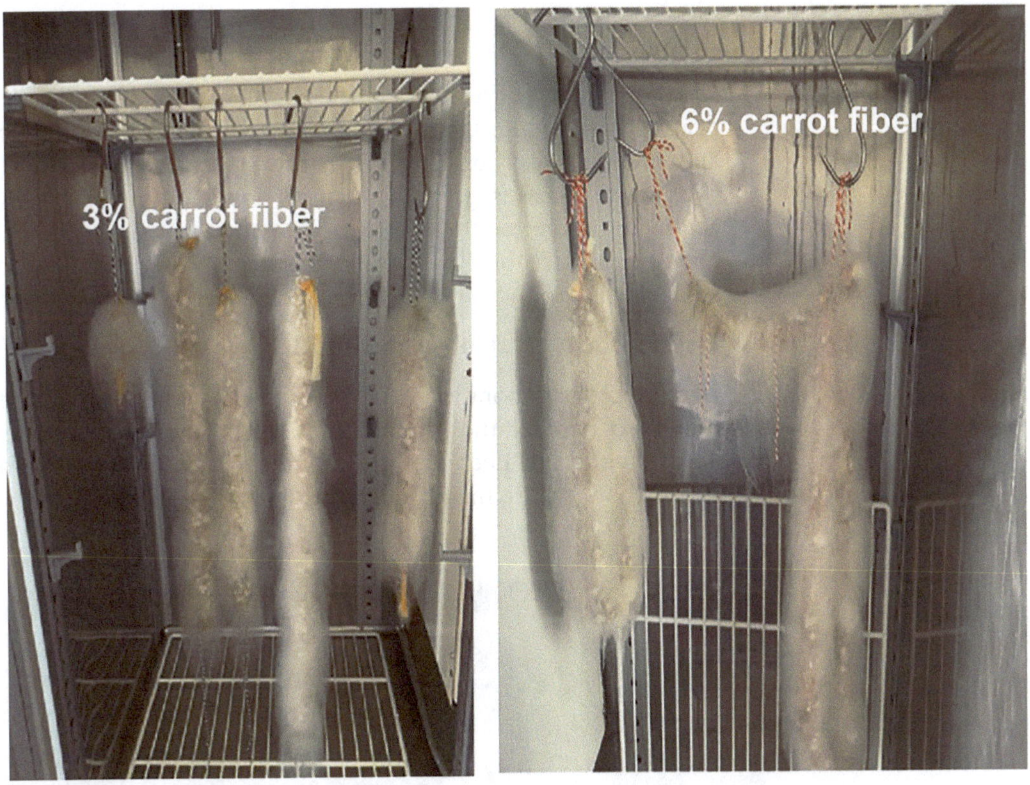

Fig. 6 Undesirable fungal growth in salchichón samples containing previously moisturized carrot fiber at 3%–6% before the addition to mixing step

take more time and high energy consumption. Thus, increased drying time induces tough texture and off-flavor [27], and could deteriorate the final product quality (tenderness, color, flavor) [28].

26. Water moves in several ways and three main transfer mechanisms are considered: (1) capillary flow of free water; (2) bound water movement; (3) diffusion in the liquid phase. These mechanisms are often competitive [29], and the drying kinetics are the consequence of both their interaction and the external conditions applied [30]. If DF and/or IFP are added, the relationship between salchichón, meat batter, and fiber changes, and several changes could affect the water loss during processing. In this aspect, high WHC are probably not suitable for this type of fermented dry-cured sausages.

27. During the ripening process, water is totally forced through convection (although the chamber has an alternation of natural and forced convection), with average values of curing air conditions over a given period, provides satisfactory prediction

results. If it is not all salchichón surface water, a hybrid system (using supersaturated solutions) can be applied (Fig. 7)

28. The consistency of fermented sausages increases due to acidification and drying. Drying is a major factor affecting binding and rheological properties. The more rapid the pH declines, the firmer the sausage is. Therefore, the addition of rapidly acidifying starter cultures and the presence of microelements promoting microbial development, such as Mn^{+2} [31] could shorten the production time because they harden the sausages.

Physicochemical changes during elaboration process

29. In salchichón elaboration process, the pH is a very important hurdle to stabilize the product and affects flavor and texture. The faster the pH decreases, the shorter the drying process can be. Carbohydrates, mainly dextrose, are the substrates used by

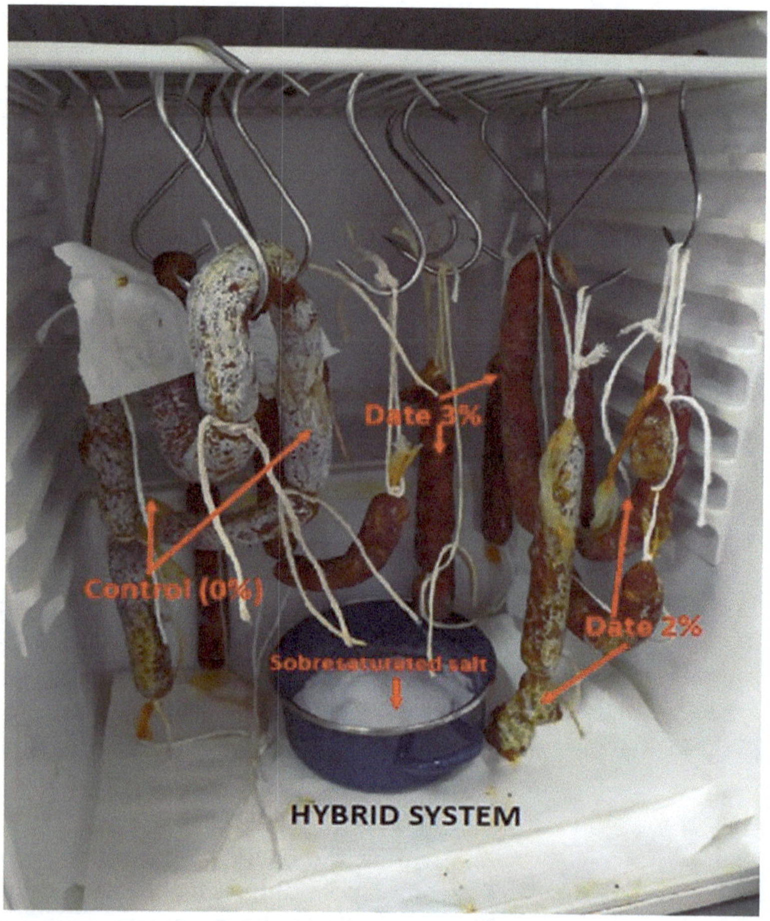

Fig. 7 Effect on salchichón of different concentrations (0%, 2%, and 3%) of fresh date IFP and hybrid system dehydration system upon shrinkage and mould growth

lactic acid bacteria (LAB), added or not as starter cultures, to produce organic acids (e.g., lactic acid). Date IFP improve this aspect. According to pH value, salchichón prototype added with DF and/or IFP can be considered as a slightly fermented dry-cured sausage if it reaches a relatively high pH (>5.3) or a highly fermented one if pH is <5.3 [32]. In Fig. 8, pH evolution during salchichón prototype added with pomegranate and date IFP elaboration process is shown.

30. Some ingredients help to speed up the fermentation process. Moreover, the Mn^{+2} present in some IFP (fresh dates) or spices accelerates pH drop and stimulates the growth of lactobacilli [31]. The magnitude and perseverance of the stimulating effect produced by Mn^{+2} differ with the type of LAB starter culture [33]. Relatively, a high level of manganese ions for optimal *L. plantarum* growth is required. Manganese is an essential element in the LAB growth. This ion is involved in the composition of some enzymes (lactate dehydrogenase) and promotes the production of enzymes that assist cellular metabolism. Mn ions (+2) are indispensable to the stability and function of the metal proteins, affecting their biological function [34].
31. Water activity: Dry-cured fermented sausages with a small caliber (initial ∅ <20 mm) have a high surface/volume relationship that facilitates oxygen penetration, increases redox potential, and hinders lactic acid bacteria (LAB) growth and pH decline. In these types of meat products, the lower effectiveness could be compensated by the addition of organic acids

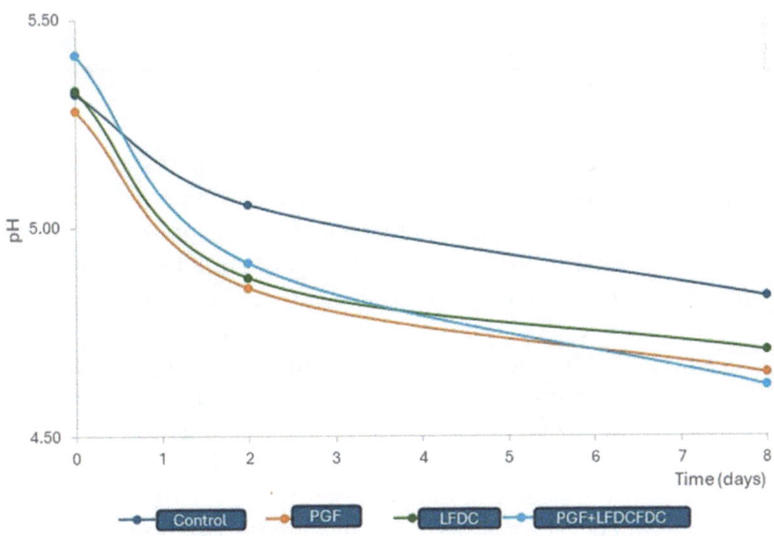

Fig. 8 Evolution of pH in salchichón prototypes (33 mm diameter) formulated with a mixture of pomegranate IFP powder [PGF], liquid date IFP [LFDC]), and the mixture of both (PGF+LFDC) during the elaboration process (8 days)

and selected starter cultures, a decrease of a_w and high-pressure treatment at the end of the process [32].

32. Moisture removal takes place through water evaporation from the outer surface. Consequently, a gradient concentration of water between the inner part of the solid and its external surface establishes.

    Microbial activity during the elaboration process.

33. During the ripening stage, the sausage surface can be sprayed with traditional microbial strains that enable the development of flavor and texture or probiotic microbiota. The microbial inoculum can be either a pure culture or a consortium of organisms [35]. The use of regenerated cellulose casing could be not suitable for salchichón DF and/or IFP added since this material is susceptible to degradation due to fungal interactions. Mould growth in dry-cured fermented sausage produces low levels of cellulases and xylanases, which initially degrade hemicellulosic casing, making a cellulose matrix, destabilization resulting in the strength loss and casing failure [12].

5 Future Perspectives

Future research should focus on identifying new fiber sources, improving extraction technologies and understanding molecular interactions between fiber components and meat proteins. The development of symbiotic meat products combining dietary fiber and probiotic cultures also represents an emerging research area.

6 Conclusions

The addition of dietary fiber (pure or IFP) into a salchichón as a fermented dry-cured sausage is feasible, with important industrial challenges to put in markets this reformulated fermented dry-cured sausage. Salchichón is an excellent model system to study the effect of dietary-fiber addition. Also represents an important opportunity to develop healthier and more sustainable foods while preserving the sensory identity of these traditional meat products.

However, successful implementation requires a comprehensive understanding of the technological, microbiological, sensorial, cultural, and structural characteristics of these types of meat products.

The use of dietary fiber derived from agro-industrial coproducts (IFP) offers a promising strategy for combining nutritional improvement, sustainability, and technological innovation within the meat industry.

Acknowledgments

This work is part of the AGROALNEXT-059 Programme financed by MCI with Nextgeneration EU funds (PRTR.C17.I1) and the Generalitat Valenciana, with the title: "Integral valorization of traditional food resources of the Valencian Community. Development of new products for agricultural and food use based on dates from Elche and Programa Iberoamericano de Ciencia y Tecnología para el Desarrollo (CYTED) through Red AlProSos 125RT0165. Cátedra Palmeral d´Elx. This study was supported by the Spanish Ministry of Science and Innovation project (PID2021-123628OB-C41-Agri-food coproducts as a source of bioactive compounds for the development of functional meat products), which is funded by MCIN/AEI/10.13039/501100011033/ and FEDER, UE. José M. Lorenzo is a member of the Competitive Reference Group "FunFood" (Functional Food), funded by GAIN (grant number IN607A2023/01).

References

1. Fernández-López J, Sendra E, Sayas-Barberá E, Navarro C, Pérez-Álvarez JA (2008) Physico-chemical and microbiological profiles of salchichón enriched with orange fiber. Meat Sci 80(2):410–417
2. López-Marcos MC, Bailina C, Viuda-Martos M, Pérez-Álvarez JA, Fernández-López J (2015) Properties of dietary fibers from agroindustrial coproducts as source for fiber-enriched foods. Food Bioprocess Technol 8:2400–2408
3. Jenkins DJA, Marchie A, Augustin LVA, Ros E, Kendall CWR (2004) Viscous dietary fibre and metabolic effects. Clin Nutr 1(2):39–49
4. López-Vargas JH, Fernández-López J, Pérez-Álvarez JA, Viuda-Martos M (2013) Chemical, physico-chemical, technological, antibacterial, and antioxidant properties of dietary fiber powder obtained from yellow passion fruit co-products. Food Res Int 51:756–763
5. Delgado-Ospina J, Martuscelli M, Grande-Tovar CD, Lucas-González R, Molina-Hernández JB, Viuda-Martos M, Fernández-López J, Pérez-Álvarez JÁ, Chaves-López C (2021) Cacao pod husk flour as an ingredient for reformulating frankfurters: effects on quality properties. Foods 10:1243
6. Lorenzo JM, Temperán S, Bermúdez R, Cobas C, Purriños L (2012) Changes in physico-chemical, microbiological, textural and sensory attributes during ripening of dry-cured foal salchichón. Meat Sci 90(1):194–198
7. Sánchez-Zapata E, Fuentes-Zaragoza E, Fernandez-Lopez J, Sendra E, Sayas E, Navarro C, Perez-Alvarez JA (2009) Preparation of dietary fiber powder from tiger nut milk by-products and its physicochemical properties. J Agric Food Chem 57(17):7725–7730
8. Sánchez-Zapata E, Díaz-Vela J, Pérez-Chabela ML, Pérez-Álvarez JA, Fernández-López J (2013) Evaluation of the effect of tiger nut fibre as a carrier of unsaturated fatty acids rich oil on the quality of dry-cured sausages. Food Bioprocess Technol 6:1181–1190
9. Sharefiabadi E, Serdaroglu M (2021) Pectin: properties and utilization in meat products. Food Health 7(1):64–71
10. Ahmad T, Sayyaf AR, Talukder S, Kumar D, Sen AR, Devadason IP (2024) Herbs, spices and condiments for processed meat products. In: Rather SA, Masoodi FA (eds) Handbook of processed functional meat products. Springer, Cham, pp 97–111
11. Sanlier N, Gencer F (2020) Role of spices in the treatment of diabetes mellitus: a mini review. Trends Food Sci Technol 99:441–449
12. Sreenath HK, Jeffries TW (2011) Interactions of fungi from fermented sausage with regenerated cellulose casings. J Ind Microbiol Biotechnol 38(11):1793–1802
13. Djordjevic J, Pecanac B, Todorovic M, Dokmanovic M, Glamoclija N, Tadic V, Baltic MZ (2015) Fermented sausage casings. Procedia Food Sci 5:69–72
14. Yan X, Yang L, Zhang Y, Han W, Duan Y (2022) Effect of collagen casing on the quality

characteristics of fermented sausage. PLoS One 17(2):e0263389
15. Marcos B, Gou P, Arnau J, Guàrdia MD, Comaposada J (2020) Co-extruded alginate as an alternative to collagen casings in dry-fermented sausages. Meat Sci 169:108184
16. Liu Y, Gao S, Cui Y, Wang L, Duan J, Yang X, Liu X, Zhang S, Sun B, Yu H, Gao X (2024) Characteristics of lactic acid bacteria as probiotic starters in fermented sausages. Foods 13(2):198
17. El-Fakhrany AMA, Rabee LA, Farouq AM, El-Garhi HEM (2026) Development of synbiotic fermented sausage enriched with Jerusalem artichoke. Food Humanity 6:101022
18. Sayas-Barberá E, Viuda-Martos M, Fernández-López J, Pérez-Álvarez JA, Sendra E (2012) Combined use of probiotic culture and citrus fiber in sausage. Food Control 27(2):343–350
19. American Meat Science Association (2012) Meat color measurement guidelines. AMSA, Champaign. pp 45–52
20. Sánchez-Zapata E, Fuentes-Zaragoza E, Navarro-Rodríguez de Vera C, Sayas E, Sendra E, Fernandez-Lopez J, Perez-Alvarez JA (2011) Effects of tuna pâté thickness on CIELAB color. Food Control 22(8):1226–1232
21. Singleton VL, Rossi JA Jr (1965) Colorimetry of total phenolics with phosphomolybdic-phosphotungstic acid reagents. Amer J Enol Viticult 16:144-58
22. Witte VC, Krause GF, Bailey ME (1970) A new extraction method for determining TBA values. J Food Sci 35(5):582–585
23. Rosmini MR, Perlo F, Pérez-Álvarez JA, Pagán-Moreno MJ, Gago-Gago A, López-Santoveña F, Aranda-Catalá V (1996) TBA test applied to pâté. Meat Sci 42(1):103–110
24. Robertson JA, de Monredon FD, Dysseler P, Guillon F, Amadó R, Thibault JF (2000) Hydration properties of dietary fibre. LWT Food Sci Technol 33:72–79
25. Cascone G, Getnet-Setegen H, Miccio M, Diaferia C (2025) A tool for modelling and simulation of irregular shape and shrinking salami during drying. Chem Eng Trans 43:103–108
26. Han DJ, Choi JH, Choi YS, Kim HY, Kim SY, Kim HW, Chung HK, Kim CJ (2011) Effects of konjac, isolated soy protein, and egg albumin on quality properties of semi-dried chicken jerky. Kor J Food Sci Anim Resour 31:183–190
27. Arnau J, Serra X, Comaposada J, Gou P, Garriga M (2007) Technologies to shorten the drying period of dry-cured meat products. Meat Sci 77:81–89
28. Kim DH, Kim YJ, Shin DM, Lee JH, Han SG (2022) Drying characteristics and physicochemical properties of semi-dried restructured sausage depending on initial moisture content. Food Sci Anim Resour 42(3):411–425
29. Baldini P, Castellani L, Diaferia C, Miccio M, Okasha F (2001) Issues and steps in modeling salami drying. In: Proceedings of the 5th Italian Conference on Chemical and Process Engineering, Florence, Italy, pp 879–884
30. Bessadok-Jemai A (2013) Characterizing the drying kinetics of high-water content agrofood particles exhibiting non-Fickian mass transport. Chem Eng Trans 32:1759–1764
31. Zaika LL, Kissinger JC (1984) Fermentation enhancement by spices: identification of active component. J Food Sci 49(1):5–9
32. Garriga M, Marcos B, Martín B, Veciana-Nogués MT, Bover-Cid S, Hugas M, Aymerich T (2005) Starter cultures and high-pressure processing to improve hygiene and safety of fermented sausages. J Food Prot 68(11): 2341–2348
33. Fitzpatrick JJ, Ahrens M, Smith S (2001) Effect of manganese on *Lactobacillus casei* fermentation to produce lactic acid from whey permeate. Process Biochem 36(7):671–675
34. Tong Y, Zhai Q, Wang G, Zhang Q, Liu X, Tian F, Zhao J, Zhang H, Chen W (2017) System-wide analysis of manganese starvation-induced metabolism in *Lactobacillus plantarum*. RSC Adv 7:12959
35. Bonilla-Luque OM, Valero A, Tomasello F, Cabo MI, Rodríguez-López P, Possas A (2024) Exploring microbial diversity during artisanal salchichón production: food safety in the consumer spotlight. LWT Food Sci Technol 191:115550

Chapter 13

Development of Beef Burgers with Encapsulated Prickly Pear Peel Extract

Rubén Agregán, Noemí Echegaray, Raquel Lucas-González, Aurora Cittadini, Mirian Pateiro and José Manuel Lorenzo Rodriguez

Abstract

The abuse of synthetic additives in commercial beef burger recipes, such as those used for preventing oxidation (e.g., butylated hydroxyanisole [BHA] and butylated hydroxytoluene [BHT]), has been linked to the risk of long-term deleterious disorders. At this point, the food industry and academia have recently researched the possibility of replacing artificial chemicals with other naturally present in plants. Fruit discarded materials are sources of biologically active substances, with antioxidant and antimicrobial properties, as shown by the prickly pear peels (PPPs), capable of increasing the shelf life of highly perishable food products. This chapter depicts the manufacture of a low-fat beef burger added with a PPP extract obtained by ultrasound-assisted extraction (UAE) and subsequently microencapsulated to preserve its biological properties. The innovative recipe presented throughout the following lines aimed at introducing a rich-antioxidant waste material in food processing as a potential preservative, yielding a high-value-added product. Sustainable-adhered procedures were designed and described in depth, highlighting the use of UAE for phenolic recovery and the subsequent spray-drying until powder.

Keywords Prickly pear peel extract, Natural antioxidant, Lipid oxidation, Shelf-Life, Antimicrobial

1 Introduction

For many years, the food industry has relied on synthetic additives to extend the shelf life of food products by delaying oxidation and inhibiting microbial growth, among other quality-degrading processes, with a clear impact on food safety. However, several widely used synthetic antioxidants, such as nitrate, nitrite, butylated hydroxyanisole (BHA), butylated hydroxytoluene (BHT), and propyl gallate, as well as colorants and other additive groups, have been linked to the development of certain health disorders as a result of their consumption. Due to the evidence reported by recent research on the subject, the food industry is gradually abandoning synthetic additives and moving toward a new plant-based approach [1].

José Manuel Lorenzo Rodriguez (ed.), *Meat Product Quality Improvement*, Methods and Protocols in Food Science, https://doi.org/10.1007/978-1-0716-5460-6_13,

Agricultural and agro-industrial activities generate large volumes of high-added organic waste for the food industry. Plant by-products are important sources of valuable chemical compounds, including oils, lipids, proteins, fibers, and phenolic compounds, among others. Fruit peels are discarded in large quantities, accounting for almost half the weight of the fruit and creating a major problem for the food processing industry. However, these and other plant-derived residues are important sources of bioactive chemicals with potential applications in the food industry as functional ingredients. In this regard, prickly pear peels (PPPs) contain a wide variety of antioxidant components, including phenolic compounds (e.g., malic, gallic, hydroxycitric, phloroglucinol, and protocatechuic acids) and flavonoids, thus showing potential for use in food matrices prone to oxidation, such as meat products. In this context, burgers are a very attractive and widely consumed product, which provides high-quality animal protein, but they are highly susceptible to lipid and microbiological deterioration [2, 3].

Therefore, plant-derived compounds, such as those from PPPs, might be added to a burger formulation to retard oxidation and extend shelf life, instead of using the aforementioned synthetic antioxidants. This chapter aimed to provide a standardized protocol for incorporating a prickly pear peel extract (PPPE) into a beef burger to preserve product quality. Extracting bioactives from plant waste materials has become an environmentally sustainable procedure that addresses the limitations of traditional techniques [4]. Ultrasound-assisted extraction (UAE) has emerged as a reference green technology for recovering antioxidant compounds, improving extraction performance, and better preserving biological activities [5]. Nevertheless, once added to the food matrix, deterioration induced by light, moisture, and oxygen could prematurely degrade sensitive compounds. Microencapsulation helps prevent this early degradation of functional compounds, preserving their biological activities for longer [6, 7].

Both techniques, UAE and spray-dryingmicroencapsulation, are applied to the innovative beef burger recipe presented in this chapter, contributing to the production of a high-value-added product while sustainably valorizing a fruit co-product, aligning with market trends toward producing "clean-label" foods.

2 Materials

2.1 Ingredients

Beef burgers are simple meat products that can be prepared with few ingredients beyond meat, such as water, salt (NaCl), and preservative additives. The burger meat recipe present herein uses beef, together with the aforementioned ingredients in the proportions

Table 1
Ingredients and their proportions by weight used in the preparation of beef burgers. PPPE: prickly pear peel extract

Ingredients	Proportion (% by weight)
Lean meat	86.5
Fat	5
Water	7.5
Salt (NaCl)	1
Encapsulated PPPE	0.5

shown in Table 1. Generally, the fat ratio is higher than that used in this chapter, which is 5%. However, this innovative recipe aims to reduce the fat content, producing a healthier standard burger and thereby increasing the amount of lean meat. Water is an important component of the mixture, as it helps disperse the ingredients and create a manageable meat dough, facilitating the subsequent formation of the burgers. Regarding the amount of salt, a small percentage is recommended to improve the flavor and texture of the final product and to enhance water and fat retention [8]. Other ingredients could also be used to produce a sensorially enhanced and more stable burger, such as seasonings, industrial preservatives (e.g., nitrites, sulfates, or ascorbic acid), and binders and texturizers (e.g., egg and milk). Nevertheless, the recipe presented herein focuses on developing a simple processed meat, avoiding overloading the matrix and the use of synthetic chemicals. In contrast, it is proposed to add a natural plant-derived compound, such as a PPPE encapsulated with maltodextrin, to serve as an antioxidant and prolong the shelf-life of the beef burgers.

The meat in the burgers can come from the beef loin, including the total amount of fat used in this recipe, rather than adding the fat separately as in some commercial preparations. On the other hand, PPPs can be obtained manually or mechanically, always removing as much pulp as possible, using ripe fruit whenever available.

2.2 Machinery Used During Product Development (see Notes 1 and 2)

To develop satisfactory meat products, the necessary machinery must be adequately incorporated, alongside the correct accessories. Table 2 shows the list of industrial and laboratory machinery for the manufacture of beef patties, added with encapsulated PPPE. In this burger preparation protocol, a series of machines were selected and suggested (Figs. 1 and 2), which are described below.

- Mincer: grinding the meat is the first step in preparing the burger. This process is essential to obtain a suitable mixture of particles of a specific size. In this case, it is recommended to attach an 8 mm diameter plate to a refrigerated meat mincer

Table 2
List of industrial and laboratory machinery for the manufacture of beef patties added with encapsulated prickly pear peel extract (PPPE) (*see* Notes 1 and 2)

Machinery	Model and brand
Mincer	La Minerva, Bologna, IT, Italy
Burger press	A-2000, Gaser, Girona, Spain
Heat sealer	A LARI3/Pn T-VG-R-SKIN, Ca.Ve.Co., Palazzolo, Italy
Gas injector	Viduca, Alicante, Spain
Ultrasound equipment	Hielscher UP400St, Teltow Germany
Homogenizer	IKA®-Werke GmbH & Co. KG, Staufen, Germany
Encapsulator equipment	Mini Spray Dryer B-290, BUCHI Ibérica S.L.U, Barcelona, Spain

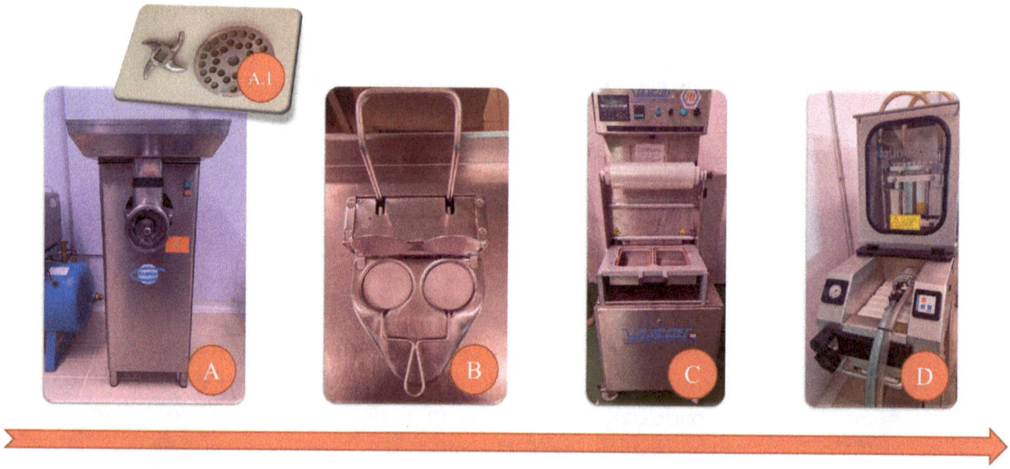

Fig. 1 Pictures of the different machines and accessories utilized in the production of beef patties. (**A**) Mincer; (**A.1**) 8 mm plate; (**B**) burger press; (**C**) heat sealer; and (**D**) gas injector. The machines are placed in order of use for the elaboration of the patties, from left to right

(La Minerva, Bologna, IT, Italy) to achieve a fine particle size suitable for proper burger formation.

- Burger press: pieces of the meat dough are molded using a burger press (A-2000, Gaser, Girona, Spain), yielding units of about 10 cm in diameter and 1 cm in length, with an approximate weight of 100 g.
- Heat sealer: considering product preservation as one of the main objectives of this recipe, the beef burgers must be packaged in a modified atmosphere and covered with a permeable film using the A LARI3/Pn T-VG-R-SKIN (Ca.Ve.Co., Palazzolo, Italy) heat sealing machine.

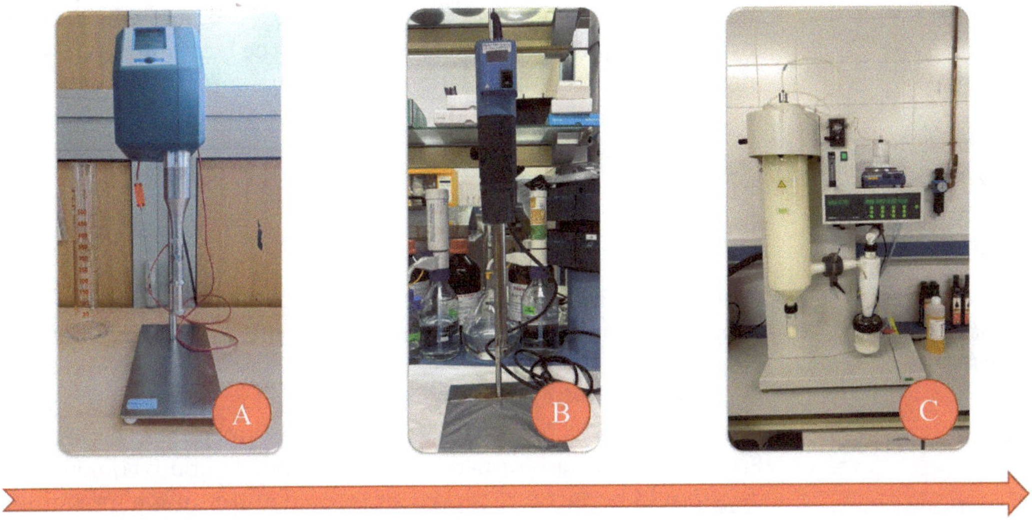

Fig. 2 Pictures of the equipment used in obtaining the encapsulated prickly pear peel extract (PPPE). (**A**) Ultrasound equipment; (**B**) Ultra-Turrax; and (**C**) spray dryer. The different pieces of equipment are placed in order of use for the production of the extract, from left to right

- Gas injector: both gases, O_2 and CO_2, are added through the permeable cover film using a Viduca (Alicante, Spain) gas injector in order to increase the product shelf life.
- Ultrasound equipment: the PPPs, previously dispersed in an ethanolic solution, are sonicated using a Hielscher UP400St ultrasonic device (Teltow, Germany), allowing the recovery of phenolic compounds and betalains to be used as antioxidants in the preparation of beef burgers.
- Homogenizer: the PPPE must be thoroughly mixed with maltodextrin, the encapsulating agent, to promote the encapsulation process using an IKA T25 digital Ultra-Turrax homogenizer (IKA®-Werke GmbH & Co. KG, Staufen, Germany).
- Encapsulator equipment: the PPPE is encapsulated in maltodextrin using the mini Spray Dryer B-290; BUCHI Ibérica S.L.U (Barcelona, Spain), which creates powdery particles.

3 Methods

The production of beef burgers with encapsulated PPPE involves several steps, some of which are crucial to obtaining a high-quality product. The recovery of the extract and its subsequent encapsulation is by far the most challenging process, presenting certain troubled points that could lead to a failed burger production. During the production of the meat product, a strict processing protocol must be followed, focusing on quality at each stage of the process, from mincing to storage. The following paragraphs describe in detail all the phases required for the manufacture of the beef burger, classified into two main sections according to the obtaining of the extract and the preparation of the product (Figs. 3 and 4).

3.1 Obtaining the Encapsulated Prickly Pear Peel Extract (PPPE)

3.1.1 Recovery of Prickly Pear Peel (PPP) Antioxidants

Peeling the prickly pear fruit is the first step to obtaining the antioxidant-rich extract. During this process, it is important to separate the pulp fraction to avoid the recovery of any type of compound other than those found in the peel. The extracted peels are then washed and dried at 45°C for 24 h until they have around 20–25% moisture. Finally, they are transformed into powder and, if not used on the same day, should be vacuum-packed and stored at freezing temperature.

- The prickly pear powder is now blended with 40% ethanol, which serves as the extraction solvent, and the mixture is subjected to sonication using the following operating parameters:
- Solid-to-solvent ratio: 1:35 (g/mL)
- Ultrasonic power: 180 W
- Frequency: 24 Hz
- Amplitude: 100%
- Limited sonication temperature: 45°C
- Time: 6 min

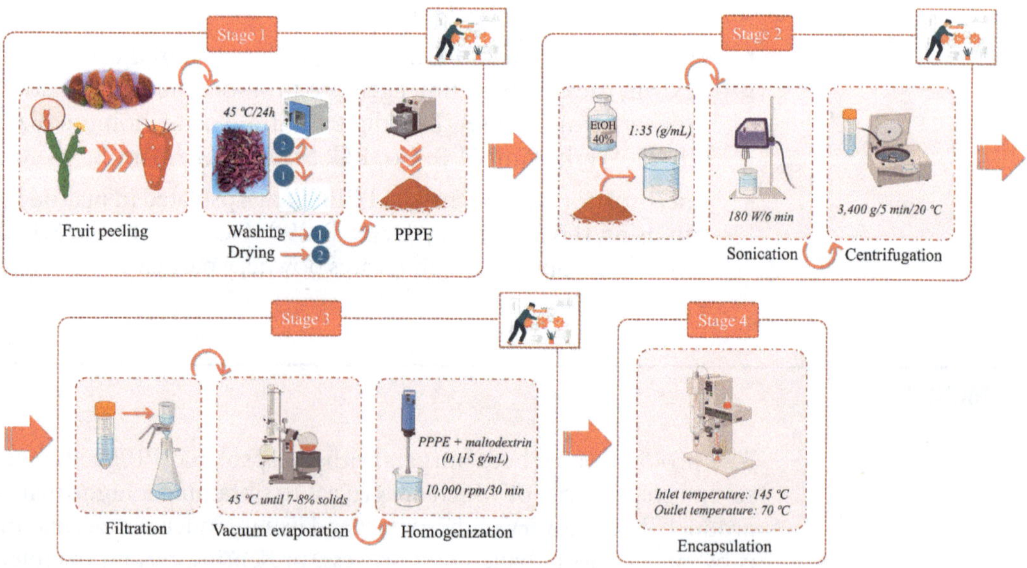

Fig. 3 Diagram of the different steps for the encapsulation process of prickly pear peel extract (PPPE). PPPE: prickly pear peel extract

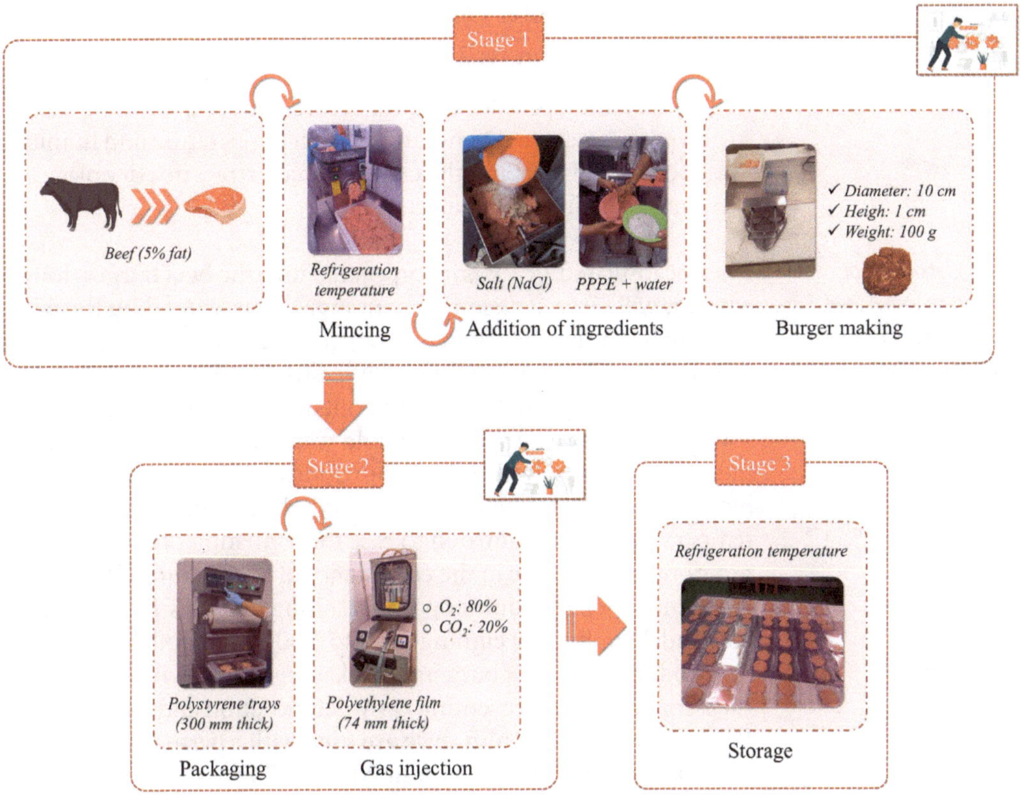

Fig. 4 Diagram of the different steps for the production of beef patties with encapsulated prickly pear peel extract (PPPE)

3.1.2 Preparation of Prickly Pear Peel Extract (PPPE) for Encapsulation

The PPPE obtained is centrifuged at 3400 × *g* for 5 min at room temperature (22 ± 1°C) to eliminate macroscopic solids, such as fibers and pectin. The resulting supernatant is subsequently collected, filtered, and finally concentrated in a rotary vacuum evaporator programmed to reach 45°C until a total solid content of 7–8% is obtained.

3.1.3 Encapsulation of Prickly Pear Peel Extract (PPPE)

Maltodextrin is the polysaccharide chosen as an encapsulating agent due to its excellent properties as a carrier, exhibiting high water solubility, low viscosity, free-flowing properties, flavor-blending ability, significant binding properties, and effective protection of bioactive compounds from oxidation [9]. The blending of this carbohydrate, resulting from the partial hydrolysis of starch [10] with PPPE, is fundamental to achieving adequate encapsulation. This step should be performed using an Ultra-Turrax homogenizer, running at 10,000 rpm for 30 min. The maltodextrin-to-extract ratio is 0.115 g/mL. The encapsulation process should be carried out under the following conditions:

- Inlet temperature: 145°C
- Outlet temperature: 70°C

The resulting powder should be stored under specific conditions to prevent premature deterioration. High temperature and humidity must be avoided, keeping the encapsulated extract in a dry place and vacuum-sealed in an opaque container.

3.1.4 Elaboration of Beef Patties with Added Encapsulated Prickly Pear Peel Extract (PPPE) (see **Note 3***)*

The encapsulated PPPE is incorporated into the beef burger dough, where it will serve as an antioxidant ingredient, extending the product's shelf life. The beef loin is chopped at refrigeration temperature to prevent early spoilage due to excessive microbial growth and is subsequently collected in a clean container. The next step is to dissolve the encapsulated PPPE in cold water and mix it well with the meat previously salted. See proportions in Table 1. The resulting meat dough should be left to rest in darkness and at refrigeration temperature for 30 or 60 min to ease the formation of the burgers by promoting an increase in the consistency of the mixture. The burgers are then shaped by a burger press, resulting in pieces 10 cm in diameter and 1 cm high, weighing around 100 g. The following step in the production of beef burgers is packaging, which will be crucial for their conversation, preventing early deterioration and loss of appeal. Polystyrene trays 300 mm thick are used with a permeable (<2 mL/(m^{2}bar/day)) polyethylene film cover of 74 mm through which a gas mixture composed of 80% O_2 and 20% CO_2 is injected. Finally, the burgers are stored in a refrigerated chamber until retailment in order to prevent premature oxidation and microbiological spoilage.

4 Notes

1. The models of the machines mentioned in this chapter are provided for guidance purposes only. The model may vary depending on the quantity of product to be manufactured. Therefore, the final decision regarding size and model is left to the reader's criteria.
2. The trademarks of the machines used in this chapter were mentioned simply as indicative examples, and their use is not mandatory. They were chosen based on the positive experience derived from their use by the authors.
3. The production of beef patties must be carried out entirely at refrigeration temperature to prevent premature microbiological deterioration.

Acknowledgments

Noemí Echegaray and Rubén Agregán acknowledge GAIN (Axencia Galega de Innovación) for their postdoctoral grants (grant numbers IN606C-2025/001 and IN606C-2025/002, respectively). Rubén Agregán, Noemi Echegaray, Mirian Pateiro, and José M. Lorenzo are members of the Competitive Reference Group "FunFood" (Functional Food), funded by GAIN (grant number IN607A2023/01). This study was supported by the Spanish Ministry of Science and Innovation project (PID2021-123628OB-C41-Agri-food co-products as a source of bioactive compounds for the development of functional meat products), which is funded by MCIN/AEI/10.13039/501100011033/ and FEDER, UE. Aurora Cittadini acknowledges the support of a "Juan de la Cierva" postdoctoral fellowship (grant number JDC2023-051424-I) funded by the Spanish "Ministerio de Ciencia, Innovación y Universidades" and the "Agencia Estatal de Investigación" MCIU/AEI/10.13039/501100011033 and the FSE+.

References

1. Nieto G, Martínez-Zamora L, Peñalver R, Marín-Iniesta F, Taboada-Rodríguez A, López-Gómez A, Martínez-Hernández GB (2023) Applications of plant bioactive compounds as replacers of synthetic additives in the food industry. Foods 13(1):47. https://doi.org/10.3390/foods130100473
2. Agregán R, Barba FJ, Gavahian M, Franco D, Khaneghah AM, Carballo J, Ferreira IC, da Silva Barretto AC, Lorenzo JM (2019) *Fucus vesiculosus* extracts as natural antioxidants for improvement of physicochemical properties and shelf life of pork patties formulated with oleogels. J Sci Food Agric 99(10):4561–4570. https://doi.org/10.1002/jsfa.9694
3. Oliveira TS, Almeida RC, de C, Silva V, de L, Ribeiro CVDM, Bezerra LR, Ferreira Ribeiro CD (2025) Enhancing beef hamburger quality: a comprehensive review of quality parameters, preservatives, and nanoencapsulation technologies of essential and edible oils. Foods 14(2):147. https://doi.org/10.3390/foods14020147
4. Usman M, Nakagawa M, Cheng S (2023) Emerging trends in green extraction techniques for bioactive natural products. Processes 11(12):3444. https://doi.org/10.3390/pr11123444
5. Moraga-Babiano L, Lucas-González R, Domínguez-Valencia R, Carrillo C, Lorenzo JM, Pateiro M (2026) Encapsulated prickly pear peel co-product extract: an antioxidant and colourant additive to preserve the quality of beef burgers packaged in modified atmosphere. Meat Sci 233:110020. https://doi.org/10.1016/j.meatsci.2025.110020
6. Jayaprakash P, Maudhuit A, Gaiani C, Desobry S (2023) Encapsulation of bioactive compounds using competitive emerging techniques: electrospraying, nano spray drying, and electrostatic spray drying. J Food Eng 339:111260. https://doi.org/10.1016/j.jfoodeng.2022.111260
7. Moraga-Babiano L, Agregán R, Echegaray N, Domínguez-Valencia R, Pateiro M, Lorenzo JM (2025) A comprehensive review of the nutritional, functional, and technological potential of prickly pear (*Opuntia ficus-indica*) in food processing. Qual Assur Saf Crops Foods 17(3):202–222. https://doi.org/10.15586/qas.v17i3.1558
8. İncesu M, Akköse A (2024) Effect of different chloride salts on chicken burger quality. Int Food Res J 31(2):331–340. https://doi.org/10.47836/ifrj.31.2.05
9. Ahmadi F, Suleria HAR, Dunshea FR (2025) Physicochemical characterization, storage stability behavior, and intestinal bioaccessibility of clove extract encapsulated using varying combinations of gum arabic and maltodextrin. Foods 14(2):237. https://doi.org/10.3390/foods14020237
10. Sahiner M, Gungor B, Silan C, Demirci S, Erdogan H, Ayyala RS, Sahiner N (2024) Synthesis and characterization of poly(maltodextrin) microgel from maltodextrin as drug delivery system. React Funct Polym 195:105826. https://doi.org/10.1016/j.reactfunctpolym.2024.105826

Chapter 14

Application of High-Power Ultrasound and Bamboo Fiber for Phosphate-Free Meat Emulsion Production

Mariana Basso Pinton, José Manuel Lorenzo Rodriguez, Bibiana Alves Dos Santos, Alexandre José Cichoski and Paulo Cezar Bastianello Campagnol

Abstract

Meat emulsions typically require phosphates to ensure stability and a desirable texture. However, excessive dietary phosphate intake has raised health concerns, including potential risks for cardiovascular and kidney diseases. Natural additives, such as dietary fibers, have been explored to reduce or eliminate phosphate in meat products. This chapter describes a detailed protocol for preparing phosphate-free meat emulsions using high-power ultrasound (HPU) and bamboo fiber (BF), a strategy that enhances emulsion stability, texture, and cohesiveness. Bamboo fiber is known for its excellent water and fat retention properties, while HPU technology promotes homogenization and modifies protein structures, significantly improving the functional properties of the final product. This combined approach enables the production of healthier emulsified meat products without compromising their technological quality, texture, or overall acceptability.

Keywords Meat emulsions, High-power ultrasound, Bamboo fiber, Phosphate replacer, Emulsion stability

1 Introduction

Phosphates are extensively used in meat emulsions to enhance water retention, emulsion stability, and texture. They also facilitate the extraction and solubilization of myofibrillar proteins, thus improving the overall quality of meat products [1, 2]. Despite these advantages, the excessive dietary intake of phosphates is associated with adverse health effects, such as increased risks of cardiovascular diseases and kidney disorders [3, 4]. Consequently, regulatory authorities and consumers demand reduced phosphate levels in processed foods, prompting the meat industry to seek alternative ingredients and technologies.

Dietary fibers have emerged as promising phosphate replacers due to their ability to retain water and fat, thereby improving

José Manuel Lorenzo Rodriguez (ed.), *Meat Product Quality Improvement*, Methods and Protocols in Food Science,
https://doi.org/10.1007/978-1-0716-5460-6_14,

emulsion stability. Among dietary fibers, bamboo fiber (BF) is particularly notable for its high fiber content, excellent hydration capacity, neutral taste, and minimal impact on product color, making it suitable for meat applications [5, 6]. Concurrently, high-power ultrasound (HPU) is gaining popularity in meat processing for its ability to improve the dispersion of ingredients, enhance emulsion stability, and alter protein structures favorably [7, 8].

The integration of HPU and BF presents a viable strategy to eliminate phosphate without compromising product quality. This chapter provides an explicit, detailed, and replicable protocol for producing stable phosphate-free meat emulsions using these combined technologies.

2 Materials

2.1 Ingredients

- Lean beef (approximately 2°C)—75% (*see* **Note 1**).
- Pork backfat (approximately 2°C)—15% (*see* **Note 1**).
- Sodium chloride (NaCl)— 1.25%.
- Potassium chloride (KCl)—0.5%.
- Bamboo fiber (particle size ~200 μm)—2.5% (*see* **Note 2**).
- Iced water (approximately 0°C)—5.75% (*see* **Note 3**).
- Water-impermeable plastic casings (46 μm thick, 32 mm diameter) (*see* **Note 4**).

2.2 Equipment

- Meat grinder equipped with a 5-mm plate.
- Bowl cutter or food processor suitable for emulsification.
- Ultrasonic bath (25 kHz, 230 W).
- Cooking water bath (capable of maintaining 80°C).
- Thermometer (probe type for internal temperature measurement).
- Vacuum packaging machine.
- Refrigerator (4°C).
- Ice bath container.

3 Methods

3.1 Preparation of Meat Emulsion

1. Grind pork backfat and lean beef separately using a meat grinder with a 5-mm plate (*see* **Note 5**).
2. Transfer the ground meat to a cutter and add NaCl, KCl, and one-third of the iced water. Emulsify (*see* **Note 6**).
3. Add bamboo fiber and the second third of iced water; emulsify again.
4. Add the ground pork backfat and the remaining iced water. Complete the final emulsification step. Ensure the emulsion temperature remains below 12°C throughout (*see* **Note 7**).
5. Fill casings with approximately 150 g of emulsion each (10 cm in length) (*see* **Note 8**).

3.2 Application of High-Power Ultrasound (Fig. 1)

1. Fully submerge the filled casings in an ultrasonic bath, maintaining water at ~20°C with ice as needed (*see* **Note 9**).
2. Apply continuous sonication (25 kHz, 230 W) for 27 min (*see* **Note 10**).

3.3 Cooking, Packaging, and Cooling

1. After sonication, cook samples in a water bath preheated to 80°C until reaching an internal temperature of 72°C (*see* **Note 11**).
2. Cool immediately in an ice bath until samples are below 10°C (*see* **Note 12**).
3. Vacuum-pack the cooled emulsions and store at 4°C.

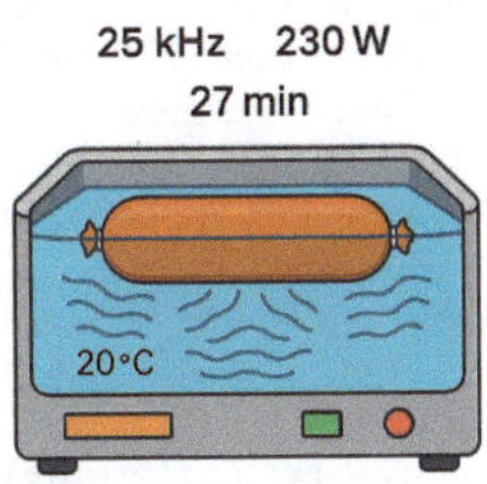

Fig. 1 Schematic representation of the ultrasonic treatment applied to phosphate-free mortadella

4 Notes

1. Maintain raw materials at ~2°C to ensure emulsion stability.
2. Control the bamboo fiber level to avoid texture defects.
3. Use accurately weighed and chilled water (ice flakes or crushed ice preferred).
4. Prepare and cut casings beforehand to ensure uniform sample dimensions.
5. A 5-mm grinding plate provides an ideal particle size for meat emulsions.
6. Emulsification should be performed in three phases, adjusting duration based on equipment.
7. Avoid temperatures >12°C to prevent fat separation and protein denaturation.
8. Remove air pockets when stuffing casings to prevent defects.
9. Use ice to maintain bath temperature and avoid heating due to ultrasonic energy.
10. Consistent sonication promotes protein modification and uniform emulsification.
11. Ensure internal temperature reaches 72°C for microbial safety.
12. Rapid cooling is essential to maintain product safety and quality.

References

1. Glorieux S, Goemaere O, Steen L, Fraeye I (2017) Phosphate reduction in emulsified meat products: impact of phosphate type and dosage on quality characteristics. Food Technol Biotechnol 55(3):390–397. https://doi.org/10.17113/FTB.55.03.17.5089
2. Goemaere O, Glorieux S, Govaert M et al (2021) Phosphate elimination in emulsified meat products: impact of protein-based ingredients on quality characteristics. Foods 10:882. https://doi.org/10.3390/FOODS10040882
3. Tentori F, Blayney MJ, Albert JM et al (2008) Mortality risk for dialysis patients with different levels of serum calcium, phosphorus, and PTH: the Dialysis Outcomes and Practice Patterns Study (DOPPS). Am J Kidney Dis 52:519–530. https://doi.org/10.1053/j.ajkd.2008.03.020
4. Younes M, Aquilina G, Castle L et al (2019) Re-evaluation of phosphoric acid–phosphates–di-, tri- and polyphosphates (E 338–341, E 343, E 450–452) as food additives and the safety of proposed extension of use. EFSA J 17(6):e05674. https://doi.org/10.2903/j.efsa.2019.5674
5. Thomas R, Jebin N, Barman K, Das A (2014) Quality and shelf life evaluation of pork nuggets incorporated with fermented bamboo shoot (*Bambusa polymorpha*) mince. Meat Sci 96:1210–1218. https://doi.org/10.1016/J.MEATSCI.2013.10.035
6. Magalhães IMC, Paglarini C, de S, Vidal VAS, Pollonio MAR (2020) Bamboo fiber improves the functional properties of reduced salt and phosphate-free Bologna sausage. J Food Process Preserv 44:e14929. https://doi.org/10.1111/JFPP.14929
7. Pinton MB, Correa LP, Facchi MMX et al (2019) Ultrasound: a new approach to reduce phosphate content of meat emulsions. Meat Sci 152:88–95. https://doi.org/10.1016/j.meatsci.2019.02.010
8. Alarcon-Rojo AD, Carrillo-Lopez LM, Reyes-Villagrana R et al (2019) Ultrasound and meat quality: a review. Ultrason Sonochem 55:369–382. https://doi.org/10.1016/J.ULTSONCH.2018.09.016

Check for updates

Chapter 15

Enhancing Hybrid Burger Quality: The Impact of Pitaya and Açaí Extracts on the Physicochemical Characteristics

Danubia Perpétua Maioto de Souza, Gabriela Xavier Oioli and Andrea Carla da Silva Barretto

Abstract

Hybrid burgers represent a viable flexitarian option, offering an alternative that can reduce ecological impacts while promoting a more balanced use of natural resources. This study evaluated the effects of pitaya and açaí extracts on the physicochemical properties of beef and hybrid burgers (with partial meat replacement by concentrated soy protein) during frozen storage. Pitaya extract contributed to the preservation of red color, whereas açaí extract exhibited fewer beneficial effects. The inclusion of soy protein improved cooking yield, reduced diameter shrinkage, and increased pH. No significant differences were observed in chewability; however, hybrid burgers displayed distinct color and texture profiles compared with meat-only formulations.

Keywords Sustainability, *Hylocereus undatus*, *Euterpe oleracea*, Reformulation, Flexitarian, Frozen storage, Color stability

1 Introduction

The significant growth of the global population has intensified the demand for food, placing significant pressure on the food industry and overburdening the use of natural resources [1]. In response, the United Nations (UN), through the Sustainable Development Goals (SDGs), emphasizes the need to transform food production and consumption systems, promoting more sustainable alternatives, such as plant-based proteins and the reduction of food waste [2].

To meet the growing consumer demand for more sustainable and healthier foods, the industry is working to reformulate traditional products [3]. One promising approach is the development of hybrid products that combine animal- and plant-based proteins in various proportions [4, 5]. These products have gained popularity in the food market as a nutritious option for consumers who wish to

José Manuel Lorenzo Rodriguez (ed.), *Meat Product Quality Improvement*, Methods and Protocols in Food Science, https://doi.org/10.1007/978-1-0716-5460-6_15,

reduce meat consumption without fully eliminating it, making them a more viable and attractive alternative to strict vegetarianism [6, 7].

The addition of soy concentrate protein in burgers has been widely reported in the literature for its positive impact on textural properties, such as hardness and cohesiveness. However, appearance attributes, such as color and shape, may be negatively affected. In this context, the use of fruit extracts, such as pitaya and açaí, has shown promise in improving the physicochemical properties of burgers. This study aims to evaluate the influence of these extracts on key characteristics, such as color, texture, yield, and pH, in both beef and hybrid burgers, providing an analysis of the effects these ingredients may have on the final product quality.

2 Materials

- Buffer solutions:
  - pH 4, pH 7 (used for pH meter calibration).
- Equipments:
  - mPA-210 digital pH meter (Tecnopon, São Paulo, Brazil).
  - ColorFlex45/0 colorimeter (Hunterlab, Reston, USA).
  - TA-XTPlus texture analyzer (Stable Micro Systems, Godalming, UK).
  - 502.150 BL digital caliper (Digimess, São Paulo, Brazil).
  - BG 2000 semi-analytical balance (Gehaka, São Paulo, Brazil).
  - Universal Software (version 4.10, Hunterlab) for color analysis.
  - Texture Exponent 32 software (version 1.0.0.68, Stable Micro Systems).
  - Freeze dryer (Liotop L108, São Carlos, Brazil).
  - Energy Pulse System (Epulsus – PM1–10, Lisbon, Portugal).
- Burger Ingredients:
  - Pork back fat and beef chuck.
  - Sodium chloride.
  - Onion powder.
  - Garlic powder.
  - White pepper.
  - Nutmeg.

– Concentrated soy protein (Arcon T 158 411, 6% moisture, 71.9% protein, Archer-Daniels-Midland Co., Chicago, USA).

3 Methods

3.1 Pitaya and Açaí Extracts

The Pitaya extract was obtained by dissolving 5 g of freeze-dried pitaya pulp in 75 mL of water, which was stirred for 10 min at 50 rpm. The PEF was applied at 5000 V, 10 Hz, and 200 pulses for 20 s. The extract was then stirred for 120 min at 150 rpm, centrifuged at 4500 rpm for 10 min, filtered, and freeze-dried. For the açaí extract, water was added to the pulp in a 1:1 ratio, stirred for 30 min at 150 rpm, and then centrifuged at 4500 rpm for 20 min. The supernatant was filtered, frozen, and freeze-dried.

3.2 Burger Formulation

The burgers were prepared in two independent replicates. A total of eight treatments were formulated: CBB—beef burger without antioxidant; EBB—beef burger with sodium erythorbate (0.05%); PBB—beef burger with pitaya extract (0.05%); ABB—beef burger with açaí extract (0.05%); CHB—hybrid burger without antioxidant; EHB—hybrid burger with sodium erythorbate (0.05%); PHB—hybrid burger with pitaya extract (0.05%); AHB—hybrid burger with açaí extract (0.05%). All the burgers were made with: 1.5% sodium chloride, 15% pork back fat, onion powder 0.6%, garlic powder 0.6%, white pepper 0.15%, and nutmeg 0.15%. The remaining ingredients used in the formulation are described in Table 1. The concentrated

Table 1
Ingredients proportion of the different burger formulations (%)

Proportion	CBB	EBB	PBB	ABB	CHB	EHB	PHB	AHB
Beef chuck	72	72	72	72	36	36	36	36
Water/Ice	10	9.95	9.95	9.95	10	9.95	9.95	9.95
Pitaya extract	—	–	0.05	–	–	–	0.05	–
Acai extract	–	–	–	0.05	–	–	–	0.05
Sodium erythorbate	–	0.05	–	–	–	0.05	–	–
Hydrated concentrated soy protein[a]	–	–	–	–	36	36	36	36

[a]1:2.18 (1 part of ARCON T ADM Protein 70% protein to 2.18 parts of water). CBB—beef burger without antioxidant; EBB—beef burger with sodium erythorbate (500 mg/kg); PBB—beef burger with pitaya extract (500 mg/kg); ABB—beef burger with açaí extract (500 mg/kg); CHB—hybrid burger without antioxidant; EHB—hybrid burger with sodium erythorbate (500 mg/kg); PHB— hybrid burger with pitaya extract (500 mg/kg); AHB—hybrid burger with açaí extract (500 mg/kg)

soy protein (Arcon T 158 411) used in the burger formulation had a moisture content of 6% and a protein content of 71.9% (*see* **Note 1**). After production, the products were stored at −18°C for 90 days and evaluated at 0, 30, 60, and 90 days.

3.3 pH Determination and Instrumental Color Measurement

The pH was determined using a digital pH meter, previously calibrated with buffer solutions. The probe was directly inserted into the product in triplicate for pH measurements (*see* **Note 2**). The instrumental color was measured using a colorimeter with D65 illuminant settings and a 10° observer angle (*see* **Note 3**). The CIELab color system was used to measure the parameters of lightness (L*), redness (a*), and yellowness (b*). Twelve readings from each treatment were collected at equidistant points on the samples for each analysis.

3.4 Texture Profile Analysis

The texture profile analysis was conducted using a texture analyzer to measure hardness, cohesiveness, springiness, and chewiness (*see* **Note 4**). Data were obtained using Texture Exponent 32 software (version 1.0.0.68, Stable Micro Systems). Ten repetitions of each treatment were performed.

3.5 Diameter Reduction and Yield

Diameter reduction was calculated using a digital caliper by measuring the diameter of each burger before and after cooking (*see* **Note 5**). The percentage of diameter reduction was calculated using Eq. 1 [8]. The equation for calculating diameter reduction is as follows:

$$\%\text{diameter reduction} = \frac{\text{initial diameter} - \text{final diameter}}{\text{initial diameter}} \times 100 \quad (1)$$

Yield was calculated by weighing two burgers from each treatment. After baking (*see* **Note 5**), the burgers were cooled to room temperature and weighed again. The percent (%) yield was calculated using Eq. 2 [8]. The equation for calculating yield is as follows:

$$\%\text{yield} = \frac{\text{final weight (g)}}{\text{initial weight (g)}} \times 100 \quad (2)$$

3.6 Statistical Analyses

Statistical analyses were performed to assess the effects of storage times and treatments on pH, color, texture, cooking losses, and diameter reduction. Results are presented as means ± SEM. Storage times and treatments were fixed factors, while processing repetitions (on different days) were treated as random effects. Data were analyzed using ANOVA, and means were compared with the Tukey test at $p < 0.05$, using the STATISTICA software version 7.0 from Statsoft Inc.

4 Notes

1. To minimize interference in the product matrix when replacing meat, a mixture of water and concentrated soy protein was previously prepared, considering that the moisture content of meat is significantly higher than that of concentrated soy protein; 1:2.18 (1 part of ARCON T ADM Protein 70% protein to 2.18 parts of water).
2. The pH meter was properly calibrated with buffer solutions at room temperature before conducting the measurements. The samples were measured at room temperature.
3. The accuracy of color measurements may be affected by factors such as the sample temperature at the time of analysis. The samples were thawed under refrigeration for approximately 20 h and then kept at room temperature for 10 min before analysis.
4. The samples were baked until they reached 72°C at the geometric center. After cooling to room temperature, the samples were cut into 23 mm diameter discs using a cylindrical stainless-steel cutter. A 36 mm diameter cylindrical probe was used with a compression rate of 50% and a test speed of 5 mm/s.
5. The balance and caliper were properly calibrated to avoid systematic errors in weight and diameter measurements. The burgers were baked until they reached a central temperature of 72°C and were kept at room temperature until they reached a constant temperature at the central point for evaluation.

5 Results and Discussion

5.1 pH

Table 2 shows the proximate composition of manufactured burgers. The partial substitution of meat with soy protein concentrates in the burgers contributed to the increase in pH (Table 3). On day 0, the hybrid burgers presented the highest pH values ($p < 0.05$), except for EHB, which was similar ($p > 0.05$) to CBB and EBB. During the storage period, there was a trend toward higher pH for the hybrid burgers, as shown at 90 days, where CHB and EHB had higher pH values ($p < 0.05$) than CBB and EBB. On day 0, these treatments did not differ ($p > 0.05$) from each other. Evaluating the addition of guarana extract as a strategy to extend the shelf life of pork burgers, reported that the pH of the treatments did not present significant differences at the beginning of the storage period [9].

For treatments with reduced meat, the addition of concentrated soy protein increased the pH values. The increase in pH was also

Table 2
Proximate composition of burgers (%)

Treatments	Moisture	Ash	Protein	Fat
CBB	63.90	1.72	13.52	9.44^{a}
EBB	64.84	1.85	13.87	8.55^{ab}
PBB	61.84	2.03	13.38	5.94^{cd}
ABB	62.00	2.35	14.16	8.56^{ab}
CHB	61.89	2.66	14.29	7.55^{abc}
EHB	64.96	2.79	14.59	5.23^{d}
PHB	63.53	2.74	14.52	5.54^{d}
AHB	63.40	2.78	14.67	7.17^{bcd}
SEM	0.353	0.124	0.143	0.326
p-value	0.051	n.s.	n.s.	0.000

Treatments: CBB—beef burger without antioxidant; EBB—beef burger with sodium erythorbate (500 mg/kg); PBB—beef burger with pitaya extract (500 mg/kg); ABB—beef burger with açaí extract (500 mg/kg); CHB—hybrid burger without antioxidant; EHB—hybrid burger with sodium erythorbate (500 mg/kg); PHB—hybrid burger with pitaya extract (500 mg/kg); AHB—hybrid burger with açaí extract (500 mg/kg) *SEM* standard error of the mean; *n.s.* not significant ($p > 0.05$) $^{a-d}$Means with different letters in the same column differ by the Tukey test ($p < 0.05$)

Table 3
pH of burgers during frozen storage (days)

Parameters	Treatments	0	30	60	90	SEM	*p*-value
pH	CBB	5.58^{Bbc}	5.80^{ABcd}	5.83^{Aabc}	5.68^{ABbc}	0.034	0.026
	EBB	5.53^{Cbc}	5.82^{Abcd}	5.75^{ABc}	5.59^{BCc}	0.033	0.001
	PBB	5.39^{Bc}	5.68^{Ade}	5.78^{Abc}	5.72^{Aabc}	0.034	0.000
	ABB	5.40^{Dc}	5.57^{Ce}	5.89^{Aabc}	5.73^{Babc}	0.041	0.000
	CHB	5.88Â^{a}	5.92^{abc}	5.89^{abc}	5.89^{a}	0.021	n.s.
	EHB	5.79^{Bab}	6.01^{Aa}	5.98^{ABa}	5.92^{ABa}	0.030	0.039
	PHB	5.91^{a}	6.04^{a}	5.98^{ab}	5.88^{ab}	0.028	n.s.
	AHB	5.86^{ABa}	5.99^{Aab}	5.94^{ABabc}	5.75^{Babc}	0.033	0.040
	SEM	0.035	0.026	0.019	0.021	–	–
	p-value	0.000	0.000	0.002	0.000	–	–

SEM standard error of the mean; *n.s.* not significant ($p > 0.05$) Treatments: CBB—beef burger without antioxidant; EBB—beef burger with sodium erythorbate (500 mg/kg); PBB—beef-burger with pitaya extract (500 mg/kg); ABB—beef burger with açaí extract (500 mg/kg); CHB—hybrid burger without antioxidant; EHB—hybrid burger with sodium erythorbate (500 mg/kg); PHB—hybrid burger with pitaya extract (500 mg/kg); AHB—hybrid burger with açaí extract (500 mg/kg) $^{a-e}$Means with different letters in the same column differ by the Tukey test ($p < 0.05$). $^{A-D}$ Means with different letters in the same line differ by the Tukey test ($p < 0.05$)

observed in pork burgers with 50% replacement of fat by hydrogelled flaxseed oil emulsions, added with different concentrations of isolated pea protein, mainly at the highest concentrations of pea protein, due to proteolytic reactions in the protein used [10]. This hypothesis may explain the higher pH values in the treatments containing soy protein presented in the present study.

Additionally, the replacement of 10%–40% of the meat with textured soy protein in cooked beef burgers showed a significant increase in pH [11]. The authors believe that this increase in pH, particularly in treatments with higher percentages of textured soy protein, is due to the slightly alkaline characteristic of soy protein.

5.2 Instrumental Color

The treatments with hybrid burgers presented the highest L* values (Table 4) and showed a significant difference ($p < 0.05$) compared to the treatments containing only meat. This may be due to the natural color of the concentrated soy protein being lighter than the color of the meat. The treatments did not show differences over time, except for the EBB treatment, where a reduction in L* value was observed when comparing the first and last days of storage. On Day 0, the EBB and the treatments with the addition of pitaya extract (PHB and PBB) presented the highest values for the a* value ($p < 0.05$). PHB showed a higher a* value ($p < 0.05$) throughout the storage period (30–90 days). A previous study had already shown that the use of pitaya extract is efficient in maintaining the red color in hamburgers [12]. This characteristic is desirable to consumers and relevant in purchasing decisions, as they often associate the vibrant red color with freshness, nutritional safety, and microbiological safety for consumption [13]. On the other hand, a reduction in the a* value was observed for ABB when comparing Day 0 with Day 90. The treatment without concentrated soy protein and açaí extract (ABB) did not differ from CBB, except at the last point of analysis (Day 90), when CBB showed a higher a* value. A previous study on the addition of açaí pulp extract in different concentrations (750 and 500 ppm) in pork burgers stored under refrigeration observed a reduction in the a* value during the first 5 days of storage [14]. However, at the end of the storage period, the açaí extract maintained the intensity of the red color when compared to the control treatment. Açaí is a fruit that contains anthocyanins, which are compounds whose color stability can be affected by variables such as pH, light, metal ions, temperature, and oxygen, among others [15]. This information suggests that the açaí extract used in the present study did not perform as well as in other studies in the literature, since the anthocyanins may have suffered degrading effects from some of the aforementioned factors. The hybrid burgers presented higher b* values compared to the treatments with meat only. This is probably due to the yellow color of the soy protein. Another similar study [14] described that increasing the concentration of açaí extract decreased b* values. The

Table 4
Instrumental color of burgers

Parameters	Treatments	0	30	60	90	SEM	*p*-value
L*	CBB	47.44^{cd}	47.13^{cd}	46.20^{c}	46.54^{bc}	0.404	n.s.
	EBB	46.08^{Ad}	43.36^{ABe}	43.63^{ABc}	42.91^{Bd}	0.401	0.022
	PBB	45.68^{de}	44.36^{de}	45.26^{c}	45.15^{cd}	0.395	n.s.
	ABB	43.04^{e}	41.48^{e}	43.46^{c}	42.69^{d}	0.342	n.s.
	CHB	53.30^{a}	52.71^{a}	52.79^{a}	52.43^{a}	0.302	n.s.
	EHB	52.26^{ab}	51.71^{ab}	51.22^{ab}	51.99^{a}	0.290	n.s.
	PHB	51.83^{ab}	50.81^{ab}	52.92^{a}	52.72^{a}	0.311	n.s.
	AHB	49.73^{bc}	48.83^{bc}	49.57^{b}	48.57^{b}	0.305	n.s.
	SEM	0.336	0.369	0.364	0.376	–	–
	p-value	0.00	0.00	0.00	0.00	–	–
a*	CBB	8.84^{Abc}	5.56^{Bcde}	5.19^{Bcd}	4.89^{Bc}	0.195	0.000
	EBB	10.36^{Aa}	6.16^{Bbcd}	5.95^{Bbc}	5.88^{Bb}	0.245	0.000
	PBB	9.40^{Aab}	6.91^{Bb}	6.41^{BCb}	5.91^{Cb}	0.180	0.000
	ABB	8.22^{Acd}	5.34^{Bde}	4.61^{BCd}	3.97^{Cd}	0.202	0.000
	CHB	7.75^{Ad}	6.73^{Bb}	6.38^{Bb}	6.19^{Bb}	0.106	0.000
	EHB	8.24^{Acd}	6.26^{Bbc}	6.15^{Bb}	5.86^{Bb}	0.129	0.000
	PHB	9.47^{Aab}	7.96^{Ba}	7.39^{BCa}	7.13^{Ca}	0.132	0.000
	AHB	5.55^{Ae}	4.95^{Be}	4.29^{Cd}	4.77^{Bc}	0.078	0.000
	SEM	0.130	0.094	0.104	0.093	–	–
	p-value	0.00	0.00	0.00	0.00	–	–
b*	CBB	15.02^{d}	14.46^{d}	14.54^{c}	14.27^{d}	0.146	n.s.
	EBB	14.29^{de}	13.82^{d}	13.69^{c}	13.54^{d}	0.131	n.s.
	PBB	13.42^{ef}	13.65^{d}	13.47^{cd}	13.37^{d}	0.129	n.s.
	ABB	12.55^{f}	12.40^{e}	12.54^{d}	12.03^{e}	0.129	n.s.
	CHB	19.81^{a}	19.25^{a}	19.02^{a}	19.46^{a}	0.128	n.s.
	EHB	19.48^{a}	18.93^{ab}	19.18^{a}	19.07^{ab}	0.138	n.s.
	PHB	18.09^{b}	17.99^{b}	18.50^{a}	17.97^{b}	0.144	n.s.
	AHB	16.33^{c}	16.06^{c}	15.78^{b}	15.96^{c}	0.107	n.s.
	SEM	0.208	0.199	0.204	0.215	–	–
	p-value	0.00	0.00	0.00	0.00	–	–

SEM standard error of the mean; *n.s.* not significant ($p > 0.05$) Treatments: CBB—beef burger without antioxidant; EBB—beef burger with sodium erythorbate (500 mg/kg); PBB—beef burger with pitaya extract (500 mg/kg); ABB—beef burger with açaí extract (500 mg/kg); CHB—hybrid burger without antioxidant; EHB—hybrid burger with sodium erythorbate (500 mg/kg); PHB—hybrid burger with pitaya extract (500 mg/kg); AHB—hybrid burger with açaí extract (500 mg/kg)$^{a-f}$Means with different letters in the same column differ by the Tukey test ($p < 0.05$). $^{A-C}$Mean with different letters in the same line differ by the Tukey test ($p < 0.05$)

increase in the concentration of black rice extract in beef burgers was inversely proportional to the reduction in the intensity of the yellow coloration [16]. In the present study, when comparing the treatments at each period, the lowest b* values averaged were observed in the treatment that contained only meat and açaí extract (ABB). This is due to the naturally violet or dark purple coloration of the açaí extract. The b* value was not influenced ($p > 0.50$) by the storage period.

5.3 Texture Profile Analysis

Texture is one of the main sensory parameters that influence consumer acceptance of meat products, and its characteristics can be affected by factors such as moisture and fat content [17]. In the CBB, EBB, and PHB treatments, the highest values of hardness and, consequently, chewiness were observed ($p < 0.05$) (Table 5). The cooking process promotes the gelation of myofibrillar proteins, which is crucial for the textural properties of the hamburger [18]. The cohesion values averaged around 0.62 N. Treatments with concentrated soy protein showed similar results ($p > 0.05$) to the treatments without meat reduction regarding cohesion, except for AHB, which showed the highest value for this parameter, with significant differences compared to the CBB and PBB treatments ($p < 0.05$).

Partial substitution of meat with concentrated soy protein does not seem to have negatively impacted chewiness, as the highest values for this parameter were observed in both meat-only and hybrid

Table 5
Yield (%), diameter reduction (%), and texture parameters of burgers (N)

Treatments	Yield	Diameter reduction	Hardness	Cohesiveness	Springiness	Chewiness
CBB	76.25^{bc}	17.17^{a}	19.11Â^{a}	0.60^{b}	0.62	7.66^{a}
EBB	73.50^{c}	18.07^{a}	17.86^{ab}	0.61^{ab}	0.61	7.15^{ab}
PBB	76.14^{bc}	20.11^{a}	14.64^{bc}	0.60^{b}	0.58	4.84^{c}
ABB	80.63^{b}	12.72^{ab}	14.63^{bc}	0.62^{ab}	0.59	4.87^{c}
CHB	91.64^{a}	6.68^{bc}	13.92^{c}	0.63^{ab}	0.65	5.37^{bc}
EHB	89.07^{a}	7.38^{bc}	14.35^{bc}	0.64^{ab}	0.63	5.74^{abc}
PHB	89.03^{a}	6.63^{bc}	16.73^{abc}	0.64^{ab}	0.62	7.02^{ab}
AHB	90.95^{a}	3.46^{c}	13.84^{c}	0.65^{a}	0.62	5.42^{bc}
SEM	1.330	1.171	0.363	0.004	0.007	0.182
p-value	0.000	0.000	0.000	0.003	n.s.	0.000

SEM standard error of the mean; *n.s.* not significant ($p > 0.05$) Treatments: CBB—beef burger without antioxidant; EBB—beef burger with sodium erythorbate (500 mg/kg); PBB—beef burger with pitaya extract (500 mg/kg); ABB—beef burger with açaí extract (500 mg/kg); CHB—hybrid burger without antioxidant; EHB—hybrid burger with sodium erythorbate (500 mg/kg); PHB—hybrid burger with pitaya extract (500 mg/kg); AHB—hybrid burger with açaí extract (500 mg/kg) $^{a–c}$Mean with different letters in the same column differ by the Tukey test ($p < 0.05$). $^{A–D}$Mean with different letters in the same line differ by the Tukey test ($p < 0.05$)

treatments. The CBB, EBB, PHB, and EHB treatments exhibited the highest chewiness values, with no significant differences between them ($p > 0.05$). Additionally, no significant differences ($p > 0.05$) were observed between the treatments regarding the elasticity parameter.

The ability to replace a portion of meat with plant-based protein without compromising texture provides an appealing option to reduce the environmental impact of meat production while maintaining the experience of traditional products [19]. Therefore, positive results may encourage further studies on alternative plant proteins and different concentrations that could offer similar or even superior textural qualities.

5.4 Diameter Reduction and Yield

The concentrated soy protein contributed to a higher yield ($p < 0.05$), resulting in a smaller reduction in diameter ($p < 0.05$) in the hybrid hamburgers, except in the ABB treatment, which showed no significant difference ($p > 0.05$) in diameter reduction when compared to the hybrid hamburgers (Table 5). This phenomenon occurs due to the water absorption by concentrated soy protein, followed by a reduction in water loss during the cooking process [20].

The present study demonstrates that concentrated soy protein contributed to water and fat absorption, resulting in less weight loss during cooking in the treatments containing concentrated soy protein. An improvement in yield was also observed in hybrid burgers that incorporated white flour from *Phaseolus vulgaris* L. as a partial substitute for pork meat [21].

6 Conclusion

The addition of soy protein concentrate increased pH and cooking yield while reducing diameter shrinkage during cooking. Hybrid burger formulations exhibited higher instrumental color values, particularly L* and b*, indicating that the lighter color of soy protein influenced the overall appearance of the burgers. However, the incorporation of pitaya extract helped preserve red color intensity during storage. Texture profile analysis showed that partial replacement of meat with soy protein did not adversely affect chewiness or cohesiveness, indicating that the hybrid formulations retained desirable sensory attributes. Overall, the combined use of soy protein concentrate and pitaya extract in meat products enhances the nutritional profile, extends shelf life, and maintains favorable textural properties.

Acknowledgments

This work was supported by CAPES (Coordenação de Aperfeiçoamento de Pessoal de Nível Superior).

References

1. Clark B, Stewart GB, Panzone LA, Kyriazakis I, Frewer LJ (2017) Citizens, consumers and farm animal welfare: a meta-analysis of willingness-to-pay studies. Food Policy 68:112–127. https://doi.org/10.1016/j.foodpol.2017.01.006
2. United Nations (2015) Transforming our world: The 2030 agenda for sustainable development. https://sustainabledevelopment.un.org/post2015/transformingourworld
3. Guiné R, Florença S, Barroca M, Anjos O (2021) The duality of innovation and food development versus purely traditional foods. Trends Food Sci Technol 109:16–24. https://doi.org/10.1016/J.TIFS.2021.01.010
4. Guerrero M, Stone AK, Singh R, Lui YC, Koksel F, Nickerson MT (2025) Effect of extrusion conditions on the characteristics of texturized vegetable protein from a faba bean protein mix and its application in vegan and hybrid burgers. Foods 14:547. https://doi.org/10.3390/foods14040547
5. Grasso S, Jaworska S (2020) Part meat and part plant: are hybrid meat products fad or future. Foods 9(12):1888. https://doi.org/10.3390/foods9121888
6. Lang M (2020) Consumer acceptance of blending plant-based ingredients into traditional meat-based foods: evidence from the meat-mushroom blend. Food Qual Prefer 79:103758. https://doi.org/10.1016/j.foodqual.2019.103758
7. Elzerman JE, Hoek AC, van Boekel MAJS, Luning PA (2011) Consumer acceptance and appropriateness of meat substitutes in a meal context. Food Qual Prefer 22:233–240. https://doi.org/10.1016/j.foodqual.2010.10.006
8. Berry BW (1992) Low fat level effects on sensory, shear, cooking and chemical properties of ground beef patties. J Food Sci 57(3):537–540. https://doi.org/10.1111/j.1365-2621.1992.tb08037.x
9. Pateiro M, Vargas FC, Chincha AAIA, Sant'Ana AS, Strozzi I, Rocchetti G, Barba FJ, Domínguez R, Lucini L, do Amaral Sobral PJ, Lorenzo JM (2018) Guarana seed extracts as a useful strategy to extend the shelf life of pork patties: UHPLC-ESI/QTOF phenolic profile and impact on microbial inactivation, lipid and protein oxidation and antioxidant capacity. Food Res Int 114:55–63. https://doi.org/10.1016/j.foodres.2018.07.047
10. de Lima Guterres L, Pinton MB, dos Santos BA, Correa LP, Cordeiro MWS, Wagner R, Cichoski AJ, Lorenzo JM, Campagnol PCB (2023) Hydrogelled emulsion from linseed oil and pea protein as a strategy to produce healthier pork burgers with high technological and sensory quality. Meat Sci 195:109028. https://doi.org/10.1016/j.meatsci.2022.109028
11. Bakhsh A, Lee SJ, Lee E-Y, Hwang YH, Joo ST (2021) Characteristics of beef patties substituted by different levels of textured vegetable protein and taste traits assessed by electronic tongue system. Foods 10(11):2811. https://doi.org/10.3390/foods10112811
12. Bellucci ERB, Munekata PES, Pateiro M, Lorenzo JM, da Silva Barretto AC (2021) Red pitaya extract as natural antioxidant in pork patties with total replacement of animal fat. Meat Sci 171:108284. https://doi.org/10.1016/j.meatsci.2020.108284
13. Suman SP, Hunt MC, Nair MN, Rentfrow G (2014) Improving beef color stability: practical strategies and underlying mechanisms. Meat Sci 98(3):490–504. https://doi.org/10.1016/j.meatsci.2014.06.032
14. Bellucci ERB, dos Santos JM, Carvalho LT, Borgonovi TF, Lorenzo JM, Silva-Barretto AC (2022) Açaí extract powder as natural antioxidant on pork patties during the refrigerated storage. Meat Sci 184:108667. https://doi.org/10.1016/j.meatsci.2021.108667
15. Cavalcanti RN, Santos DT, Meireles MAA (2011) Non-thermal stabilization mechanisms of anthocyanins in model and food systems—an overview. Food Res Int 44(2):499–509. https://doi.org/10.1016/j.foodres.2010.12.007
16. Prommachart R, Belem TS, Uriyapongson S, Rayas-Duarte P, Uriyapongson J, Ramanathan R (2020) The effect of black rice water extract on surface color, lipid oxidation, microbial growth, and antioxidant activity of beef patties during chilled storage. Meat Sci 164:108091. https://doi.org/10.1016/j.meatsci.2020.108091
17. Kim TK, Shim JY, Hwang KE, Kim YB, Sung JM, Paik HD, Choi YS (2018) Effect of hydrocolloids on the quality of restructured hams

with duck skin. Poult Sci 97(12):4442–4449. https://doi.org/10.3382/ps/pey309

18. Sun XD, Holley RA (2011) Factors influencing gel formation by myofibrillar proteins in muscle foods. Compr Rev Food Sci Food Saf 10(1):33–51. https://doi.org/10.1111/j.1541-4337.2010.00137.x
19. Bis-Souza CV, Barba FJ, Lorenzo JM, Penna ALB, Barretto ACS (2019) New strategies for the development of innovative fermented meat products: a review regarding the incorporation of probiotics and dietary fibers. Food Rev Int. https://doi.org/10.1080/87559129.2019.1584816
20. Sánchez-Zapata E, Muñoz CM, Fuentes E, Fernández-López J, Sendra E, Sayas E, Pérez-Álvarez JA (2010) Effect of tiger nut fibre on quality characteristics of pork burger. Meat Sci 85(1):70–76. https://doi.org/10.1016/j.meatsci.2009.12.006
21. Argel N, Lorenzo G, Domínguez R, Fraqueza M, Fernández-López J, Sosa M, Campagnol P, Lorenzo J, Andrés S (2022) Hybrid meat products: incorporation of white bean flour in lean pork burgers. Appl Sci 12(15):7571. https://doi.org/10.3390/app12157571

Check for updates

Chapter 16

Lactiplantibacillus plantarum BFL as a Probiotic Source in *Salamines Criollos*

Noelí Estefanía Sirini, Laureano Sebastián Frizzo, Lorena Paola Soto, Jorge Alberto Zimmermann, Marcelo Lisandro Signorini, María Virginia Zbrun, Florencia Aylen Lencina and María Ángeles Stegmayer

Abstract

In Argentina, the salami production began in the mid-nineteenth century in cities where Italian immigrants had settled. *Salamin Criollo* is a traditional dry-fermented sausage manufactured in the Litoral area of Argentina and is currently consumed throughout the country. In the manufacture of *Salamines Criollos*, both pork and beef meat are chopped and mixed with salt, pork back fat, culture starters, and additives. *Salamin Criollo* has a length of 13–15 cm, a diameter of 3.5–4.5 cm, and the pork back fat is coarsely chopped. Due to dry-fermented sausages not being subjected to heat treatments, they represent a suitable matrix to transport probiotics. In addition, dry-fermented sausages as carriers of probiotics give place to an innovative alternative of new non-dairy functional foods. The aim of this chapter is to describe the formulation and the different stages of *Salamin Criollo* manufacturing as a vehicle for the probiotic *Lactiplantibacillus plantarum* BFL.

Keywords Healthy meat products, Functional foods, Probiotics, Fermented sausages, *L. plantarum*

1 Introduction

Fermentation practices are influenced not only by the characteristics of the raw materials but also by climatic conditions, as well as social, cultural, religious, and economic factors specific to the production region. As a result, fermented products are often closely linked to their geographical origin and local traditions. The preparation of dry-cured or dry-fermented sausages is part of a long-standing tradition in Argentina, where family customs and culinary knowledge come together to form a distinctive seasonal practice. Italian immigration in the nineteenth century has left a lasting influence on the typical dishes of Argentina's Litoral region [1]. This rich Italian heritage merged

José Manuel Lorenzo Rodriguez (ed.), *Meat Product Quality Improvement*, Methods and Protocols in Food Science,
https://doi.org/10.1007/978-1-0716-5460-6_16,

deeply with local Argentine traditions, giving rise to *Salamines Criollos.* According to the Argentine Food Code (Código Alimentario Argentino, CAA), the term *salamines* refers to a dry-cured sausage made from pork, or a combination of pork and beef, with the addition of pork fat, salt, saltpeter, sugar, spices, and wine [2]. In addition, the term *criollo* holds a specific historical and cultural significance in the Argentine context. It traditionally refers to individuals born in the Americas of European descent, particularly during the colonial period. In Argentina, the term was used to distinguish those born in the territory of the Viceroyalty of the Río de la Plata from those born in Spain, who were referred to as *peninsulares.*

During salami production, microorganisms naturally present in the raw materials and/or those introduced through inoculation are responsible for driving the fermentation process. The composition of the microbiota within the meat matrix is specific to each animal species, geographic region, or processing facility, depending on the microbial starter cultures used and the production practices applied. This fermentation process involves a sequence of biochemical and microbial events, and each stage—production, drying, and ripening—must be carefully monitored to ensure both hygienic safety and the desired sensory quality of the final product [3, 4].

In the case of probiotic meat products, additional technological challenges must be addressed to maintain probiotic viability throughout processing and storage. Ensuring a high survival rate of probiotic strains during production and shelf life is critical to maintaining the functional properties of the product and consumer trust. Several factors influence probiotic viability, including food-related parameters such as pH, titratable acidity, oxygen levels, water activity, sodium concentration, sugar content, and the presence of inhibitory substances such as hydrogen peroxide, bacteriocins, artificial flavorings, and colorants. Processing conditions such as heat treatment, incubation temperature, cooling rate, packaging materials, storage conditions, production scale, and microbiological variables (e.g., probiotic strain, inoculum size, and ratio) also play a significant role [4].

All of these factors must be controlled throughout production and storage to ensure that the effective probiotic dose per serving is preserved to deliver potential health benefits. In the case of the *Salamín Criollo* described in this study, the formulation was specifically developed to deliver the recommended daily probiotic dose through the consumption of just a few slices.

The technical descriptions provided in this chapter offer valuable insights for the continued development of future meat products, where the microbial intake through diet will become an increasingly relevant factor.

1.1 Salamines Criollos Characteristics

Regarding the external appearance of *Salamines Criollos*, their shape is defined by the natural bovine casing; thus, it cannot be strictly uniform. However, the weight is calibrated at approximately 250 g.

The *Salamines Criollos* color is intense red, their consistency is firm, and their aroma is intense, cured meat. Their flavor is pleasant and balanced, with notes of meat, fat, and spices. In some cases, they can be slightly acidic and spicy. Regarding texture, it is firm but not too dry, and when sliced, the bacon remains perfectly integrated into the slice (Fig. 1).

Their suitability for consumption must be verified, as indicated by:

a) A weight loss of 30% from its original wet weight,
b) The absence of off-odors and off-flavors, other than those characteristic of salami as specified,
c) Compliant microbiological and physicochemical analyses.

2 Materials

1. **Beef meat:** Lean cuts are selected from both the forequarters and hindquarters. These cuts are carefully trimmed to obtain bright red, lean meat.

   Pork meat: Lean cuts are selected from the forequarters and hindquarters. The cuts are trimmed to remove most of the visible fat.

   Pork back fat: Firm-consistency fat obtained from the dorsal area, with a white to pale pink color. It is carefully trimmed to ensure the absence of hair and skin.
2. **Probiotic inoculum** (*Lactiplantibacillus plantarum* BFL).
3. **Seasoning ingredients:** Common salt, nitrifying salt (optionally), pepper, paprika, and garlic.
4. **Casings:** Natural bovine casings of 25 mm diameter.

Fig. 1 Typical Argentinian *Salamín Criollo*

3 Methods

3.1 Salamines Criollos Manufacturing

The manufacturing process of probiotic *Salamín Criollo* consists of a series of steps, which are shown in Fig. 2. These steps include lean pork, lean beef, and pork back fat selection, mincing and mixing, probiotic inoculum and spices addition, mixing, stuffing, and drying. Each of these steps is described in detail below:

1. Probiotic inoculum production (*see* **Notes 1 and 2**).
2. Raw material control: The selection of lean beef, lean pork trimmings, and pork back fat was carried out at a maximum temperature of 5°C, with a pH range between 5.4 and 6.2.
3. Meat mincing: The lean pork and lean beef trimmings were received in frozen pieces. Both types of meat were then minced into uniform portions. The pork fat was cut into cubes

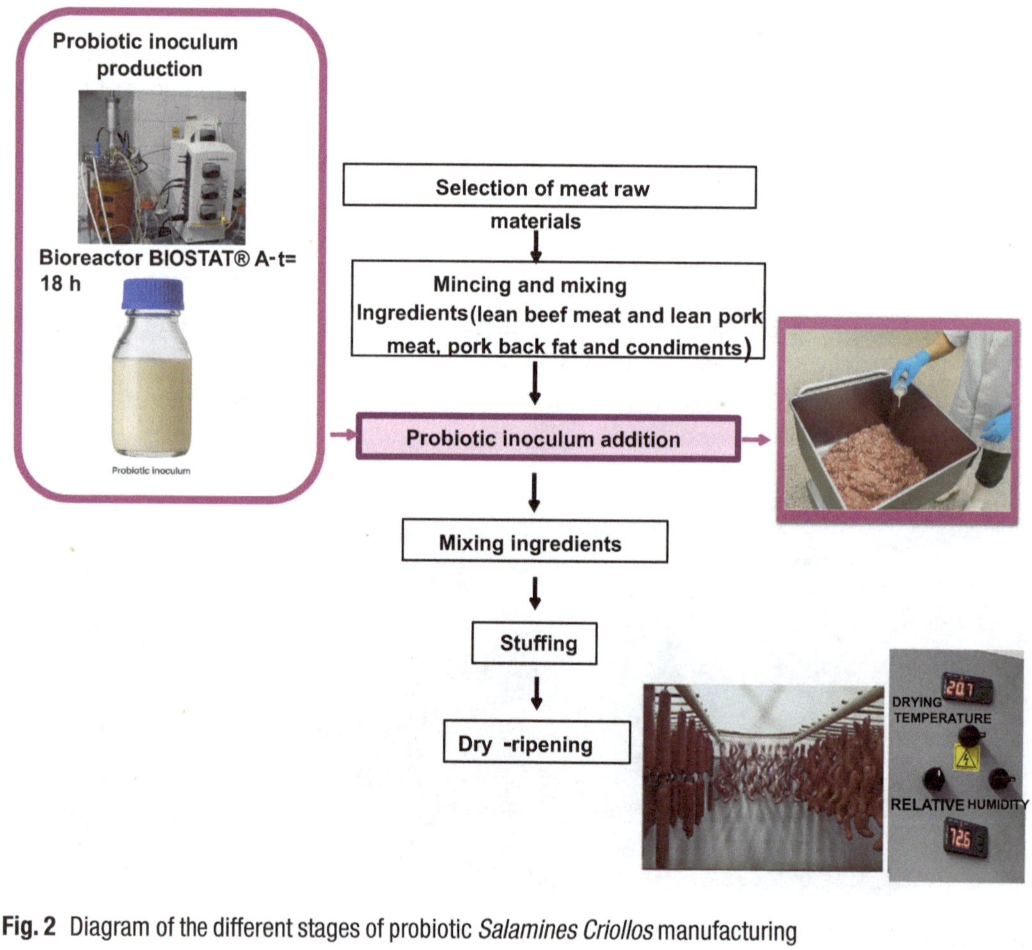

Fig. 2 Diagram of the different stages of probiotic *Salamines Criollos* manufacturing

measuring 10 mm per side using industrial equipment and then frozen for later use.

4. Ingredients mixing: Pork and beef minced meat, and pork back fat chunks were mixed according to the established formulation.
5. Addition of the probiotic inoculum, seasoning, and subsequent mixing (*see* **Note 3**).
6. Stuffing stage: The meat batter was stuffed into a natural bovine casing.
7. The drying and ripening processes were conducted for a minimum of 7 days, counted from stuffing. These processes took place in controlled drying chambers, which maintain optimal temperature and ventilation conditions to ensure appropriate levels of dry temperature and relative humidity. Drying was carried out under controlled conditions of 20°C ± 1°C and 72% ± 1% relative humidity (RH). At the end of the process, the ready-to-eat probiotic products showed viable counts of 8.8 ± 0.1 log CFU/g.

3.2 Probiotic *Salamines Criollos* Characterization

The viability of the probiotic *L. plantarum* BFL in the final product at different storage temperatures (5°C and 20°C) over a period of up to 60 days is shown in Table 1 [5]. To quantify *L. plantarum* BFL, a *Lactobacillus plantarum* selective culture medium (LPSM) was used [1, 5, 6]. The physicochemical characteristics of the final product are specified in Table 2 [1, 7]. As shown in Table 1, at both storage temperatures, the probiotic *L. plantarum* BFL demonstrated the ability to withstand the intrinsic conditions of the *Salamín Criollo* matrix (pH and water activity), maintaining viable counts above 7.5 log CFU/g over 60 days. As expected, refrigeration conditions (5°C) favored probiotic survival, reaching levels of over 8.5 log CFU/g. These findings highlight the importance of maintaining refrigerated storage during the commercialization of probiotic *Salamín Criollo* to ensure the presence of an optimal probiotic dose in a small portion of the product. Accordingly, 15 g (approximately two slices) of *Salamines Criollos* containing *L. plantarum* BFL stored at 5°C for up to 60 days could provide consumers with the recommended daily dose of probiotics to exert a beneficial health effect [8, 9].

With respect to water activity (Aw), by the end of the drying period, Aw values fell below 0.95, a threshold generally accepted as suitable for ensuring food quality. Regarding pH values, the presence of *L. plantarum* BFL in the *Salamín Criollo* formulation resulted in a decrease of approximately 0.5 pH units compared to the conventional formulation (without probiotics). This increased acidity was confirmed both instrumentally and through consumer-based sensory analysis (Table 2). However, no significant differences were found in the overall liking scores between the probiotic and conventional

Table 1
***Lactiplantibacillus plantarum* BFL viability (log UFC/g) in *Salamín Criollo* up to 60 days of storage**

Storage temperature	*L. plantarum* BFL counts (log UFC/g)
5°C	8.56
20°C	7.63

Table 2
Physicochemical characteristics of *Salamín Criollo* with *Lactiplantibacillus plantarum* BFL in the final product

Physicochemical parameters	Values
pH	4.4–4.5
Aw	0.91–0.92
Color	L*: 39–40
	a*: 17–17.5
	b*: 7–8

L*: Lightness, a*: redness, b*: yellowness

Salamines Criollos, which is consistent with findings reported by other authors who studied different types of meat products [10, 11].

According to instrumental analysis, the color of the probiotic *Salamín Criollo* remained within the expected pink range for this type of product. Moreover, consumers described the color as "pleasant," with no significant differences observed between the conventional *Salamín Criollo* and other salamines [5].

4 Notes

1. The probiotic strain *Lactiplantibacillus plantarum* BFL was isolated exclusively for research purposes from the commercial product Bioflora™ (BIOSIDUS S.A., Buenos Aires, Argentina), which is marketed in Argentina as a certified probiotic with recognized health claims. Details of the isolation procedure have been described in previously published studies [1, 5].
2. *Lactiplantibacillus plantarum* BFL inoculum production: A fresh inoculum of 13.3 ± 0.1 log CFU/mL was prepared. To produce the inoculum, a bioreactor (BIOSTAT® A Sartorius Stedim Biotech, mod. BB- 8822000, Guxhagen, Germany) was used for 18 h. The technical specification was published

by Sirini et al [5]. The biomass obtained was centrifuged, and the pellet was mixed with a 10% whey protein isolate and a maltodextrin cryoprotective solution and then preserved in an ultrafreezer at −80°C until use. Then, when the inoculum is incorporated into the meat batter, it must be completely defrosted and added in liquid form. The amount of 3.3 mL/kg meat batter of the fresh inoculum was added (Fig. 2).

3. *Salamin Criollo* formulation: Beef lean meat 50%, pork lean meat 30%, pork back fat 17%, salt 2.5%, whole black peppercorns 0.05%, $NaNO_2$ 150 ppm, probiotic inoculum 3.33 mL/kg batter. Additional condiments, such as paprika, garlic, and wine, may be incorporated based on formulation preferences [7].

Acknowledgments

This study was funded by the National Agency for Scientific and Technological Promotion (ANPCyT) [Project PICT 2022-09-00852] and the Universidad Nacional del Litoral, Santa Fe, Argentina [Projects CAI+D 85520240100084LI and CAI+D Orientado 21820210100098LI].

References

1. Sirini N, Stegmayer M, Ruiz MJ, Cuffia F, Rossler E, Otero J, Frizzo L (2023) Applicability of the probiotic Lactiplantibacillus plantarum BFL as an adjunct culture in a dry fermented sausage. Meat Sci 200:109166
2. Administración Nacional de Medicamentos, Alimentos y Tecnología Médica (ANMAT) (2023) Código Alimentario Argentino: Capítulo VI – Alimentos cárneos y afines, Artículo 339. https://www.argentina.gob.ar/anmat/codigoalimentario
3. Sirini N, Frizzo LS, Aleu G, Soto LP, Rosmini MR (2021) Use of probiotic microorganisms in the formulation of healthy meat products. Curr Opin. Food Sci 38:141–146
4. Ammor MS, Mayo B (2007) Selection criteria for lactic acid bacteria to be used as functional starter cultures in dry sausage production: an update. Meat Sci 76(1):138–146
5. Sirini N, Loyeau P, Ruiz M, Stegmayer M, Soto L, Werning M et al (2022) Development of probiotic fermented sausages and viability monitoring of supplemented Lactiplantibacillus plantarum BFL strain. Fermentation 8(10):526
6. Bujalance C, Jiménez-Valera M, Moreno E, Ruiz-Bravo A (2006) A selective differential medium for *Lactobacillus plantarum*. J Microbiol Methods 66(3):572–575
7. Sirini N, Loyeau P, Ruiz M, Stegmayer M, Soto L, Werning M et al (2022) Development of probiotic fermented sausages and viability monitoring of supplemented *Lactiplantibacillus plantarum* BFL strain. Fermentation 8(10):526
8. Turroni F, Ventura M, Butto LF, Duranti S, O'Toole PW, Motherway MOC, Van Sinderen D (2014) Molecular dialogue between the human gut microbiota and the host: a *Lactobacillus* and *Bifidobacterium* perspective. Microbial Cell Fact 13(Suppl 1):S7
9. Fontana L, Bermudez-Brito M, Plaza-Díaz J, Muñoz-Quezada S, Gil A (2013) Sources, isolation, characterization and evaluation of probiotics. BJN 109(S2):S35–S50
10. Dong C, Wang B, Li F, Zhong Q, Xia X, Kong B (2020) Effects of Edible Chitosan Coating on Harbin Red Sausage Storage Stability at Room Temperature. Meat Sci 159:107919
11. Rubio R, Jofre A, Aymerich T, Guardia MD, Garriga M (2014) Nutritionally enhanced fermented sausages as a vehicle for potential probiotic lactobacilli delivery. Meat Sci 96(2):937–942

Chapter 17

Reformulation of Dry Fermented Sausages (Chorizo) with Probiotic

Carlos Pasqualin Cavalheiro, Ana Maria Herrero and Claudia Ruiz-Capillas

Abstract

Reformulation of dry-fermented sausages (chorizo) with probiotics allows for the production of a healthier meat product for the consumer without significantly modifying the traditional characteristics of the original product. However, the quality of chorizo made with probiotics depends on the strategy used for its incorporation. This article describes both how to prepare the probiotic encapsulated in alginate pearls to enhance the viability of *Lactobacillus plantarum* throughout processing and storage, and how to incorporate it during chorizo production. This strategy was chosen because it was one of the most effective in maintaining the highest levels of probiotics and maintaining the quality of the product after processing, both technologically, sensorially, and microbiologically. The probiotic chosen was *Lactobacillus plantarum*, which has a dual role when incorporated into chorizo: on the one hand, its important function in the fermentation process, and on the other, given its effect as a probiotic and the improvement in consumer health.

Keywords Dry-fermented sausages "chorizo", Probiotic, *Lactobacillus plantarum*, Alginate beads

1 Introduction

Fermentation of meat products is a traditional method for their conservation, during which several biochemical and microbiological changes occur [1, 2]. Dry-fermented sausages are highly appreciated by consumers worldwide because of their convenience, stability, and sensory attributes [3]. Dry-fermented sausages, such as Spanish "chorizo," are one of the most popular dry-fermented meat sausages in Spain, with various regional varieties that reflect local customs and cultural diversity. These varieties differ primarily in the type (natural or artificial) and diameter of the casings, the size and shape of the sausage units, the lean-to-fat ratio in the mixture, the degree of meat mincing, the identity and quantity of added spices, and the specifics of the production process, such as smoking practices, environmental

José Manuel Lorenzo Rodriguez (ed.), *Meat Product Quality Improvement*, Methods and Protocols in Food Science,
https://doi.org/10.1007/978-1-0716-5460-6_17,

conditions, and maturation duration [4]. Dry-fermented sausages, "chorizos," consist of mincing pork and pork backfat, which are then mixed with salt and the other ingredients and condiments, with the sausage batter stuffed into natural or artificial (water-permeable) casings. After this stage, these sausages can be prepared as fresh/semidried or as dry-cured sausages. Traditionally, it is produced either with its natural flora or by adding a starter culture, mainly composed of lactic acid bacteria (LAB) and staphylococci or micrococci [5]. For this reason, in chorizos, counts greater than 10^7 CFU/g of LAB are expected at the end of ripening [6].

The fermentation induced in chorizos by LAB results in acid-fermented sausages subjected to fermentation with pH values near or lower than 5.0 and dried to an aw point that inhibits the growth of spoilage bacteria, that is, lower than 0.9 [7]. Among the LAB, *Lactobacillus plantarum* is one of the most widespread bacteria of the *Lactobacillus* genus, but there are also common probiotics from the genera *Bifidobacterium*, *Enterococcus*, and *Pediococcus* [8]. In addition, *L. plantarum* can act as a probiotic, a concept defined by the Food and Agriculture Organization/World Health Organization [9] as "live microorganisms, which, when administered in adequate amounts, confer health benefits to the consumer." These microorganisms have been applied in various types of foods, including chorizos, which may also be a useful probiotic carrier because the sausage matrix seems to protect them through the gastrointestinal tract (GIT) [10]. The addition of probiotics to meat products has been performed to improve their health implications [11].

Nevertheless, salt and nitrite contents, low pH, and water activity may reduce probiotics' viability during fermentation, ripening, and storage of products [12]. To address this problem, encapsulation, entrapment, or immobilization of probiotic bacteria has been proposed to protect cells in food and the GIT. These are methodologies whereby solids, liquids, or gaseous materials are placed in an inert shell (capsule) and are released at a controlled rate under specific conditions. The most common encapsulation techniques are extrusion, emulsification, and spray drying, with capsules ranging from a few micrometers to a few millimeters [13]. Pasqualin Cavalheiro et al. [14] assessed the use of different strategies of *L. plantarum* incorporation [free cells, alginate beads, simple (water-in-oil, W/O) and double (water-in-oil-in-water, W/O/W) emulsions] in order to maintain the probiotic viability during the ripening of chorizo and concluded that the addition of probiotics in alginate beads was the most effective strategy. Therefore, the main objective of this chapter is to provide a detailed description of the preparation of the probiotic *Lactobacillus plantarum* for its incorporation during the processing of dry-fermented chorizo sausages.

2 Materials and Procedures

2.1 Preparation of Stock Culture of Lactobacillus plantarum

Materials:

- *Lactobacillus plantarum* culture (Christian Hansen, Denmark).
- De Man, Rogosa, and Sharpe (MRS) sterile broth (Panreac Química S.A., Spain).
- Glycerol (sterile) (Panreac Química S.A., Spain).
- Analytical balance.
- 250 mL Erlenmeyer flask or sterile container.
- Manual stirrer or sterile spatula.
- Incubator (set at 37°C).
- Eppendorf tubes (1.5 or 2.0 mL, sterile).
- Micropipette and sterile tips.
- Ultrafreezer (−80°C).

Procedure (*see* **Note 1**):

1) Aseptically weigh 10 g of the *Lactobacillus plantarum* culture powder.
2) Transfer the powder to a sterile container containing 90 mL of previously sterilized MRS broth (*see* **Note 2**).
3) Manually stir until the powder is completely dissolved.
4) Incubate the mixture at 37°C for 24 h.
5) After incubation, add 20 mL of sterile glycerol to the culture and mix thoroughly to ensure homogeneity (*see* **Note 3**).
6) Aliquot 1 mL of the final mixture into sterile Eppendorf tubes (*see* **Note 4**).
7) Store the tubes in an ultrafreezer at −80°C until further use (*see* **Note 5**).

2.2 Preparation of Probiotic Culture

Materials:

- Cryopreserved Eppendorf tube containing the bacterial strain.
- De Man, Rogosa, and Sharpe (MRS) sterile broth (Panreac Química, Spain) (*see* **Note 2**).
- Vortex mixer.
- 50 mL Falcon tubes (sterile).
- Sodium chloride (Sigma-Aldrich, Spain) to prepare 0.9% M saline solution (sterile) (*see* **Note 6**).
- Centrifuge capable of reaching 5000 × *g*.
- Incubator (set at 37°C).

- Analytical balance (if needed).
- Micropipettes and sterile tips.
- Refrigerator (4°C).
- Timer or chronometer.

Procedure (*see* **Note 7**):

1) Allow one cryopreserved Eppendorf tube to thaw at room temperature for 15–20 min, or until fully liquefied.
2) Vortex the Eppendorf tube to ensure complete homogenization of the content.
3) Inoculate the entire volume of one Eppendorf tube into 100 mL of sterile MRS broth and incubate at 37°C for 24 h.
4) After incubation, transfer 35 mL of the culture to a sterile 50 mL Falcon tube.
5) Centrifuge the sample at 5000 × *g* for 20 min at 4°C. (*see* **Note 8**)
6) Carefully discard the supernatant (MRS broth) without disturbing the pellet.
7) Add 5 mL of sterile 0.1 M saline solution to the pellet and vortex or shake until the pellet is completely resuspended (*see* **Note 9**).
8) Store the resuspended cells at 4°C until further use.

2.3 Elaboration of Alginate Beads

Materials:

- Distilled water.
- Sodium alginate (Texturalia, Spain).
- Powdered milk (La Lechera, Nestlè, Spain).
- Dextrin (Cargill, Spain).
- Trehalose (Cargill, Spain).
- Probiotic culture suspension in saline solution.
- Calcium chloride (Sigma-Aldrich, Spain) to prepare 0.1 M calcium chloride solution (*see* **Note 10**).
- Separatory funnel.
- 5 mL pipette tip (200 μm diameter).
- Magnetic stirrer.
- Beaker with magnetic stir bar.
- Refrigeration system (4°C).

Procedure (*see* **Note 11**):

1) For 500 g of alginate beads. Put 423 mL of sterile distilled water in a beaker; dissolve 4.23 g of sodium alginate, 4.23 g of

powdered milk, 4.23 g of dextrin, and 21.15 g of trehalose. Stir until fully dissolved.

2) Once the solution is homogeneous, add 42.37 mL of the microbial suspension and mix gently.
3) Transfer the mixture to a separatory funnel connected to a 5 mL pipette tip (200 μm) (*see* **Note 12**). Dropwise release the mixture into a 0.1 M calcium chloride solution under continuous stirring. Maintain a 10 cm distance between the pipette tip and the calcium chloride surface to ensure proper bead formation.
4) Allow the capsules to remain in the calcium chloride solution for 30 min to complete gelation (*see* **Note 13**).
5) After the gelation period, separate the capsules from the calcium chloride solution using filtration or decantation.
6) Store the capsules (Fig. 1) under refrigeration (4°C) until further use.

2.4 Probiotic Chorizo Elaboration

Materials:

- Pork meat.
- Pork backfat.
- Choravi (ANVISA, Spain).

Fig. 1 Alginate beads containing *L. plantarum*

- Sodium chloride (NaCl).
- Curavi (ANVISA, Spain).
- *Lactobacillus plantarum* encapsulated in alginate beads (37.5 g/kg) (*see* **Note 14**).
- Meat grinder with 4.5 mm plate (MAINCA, Granollers, Spain).
- Mixer or homogenizer (Van Dall Srl, Treviglio, Italy).
- Collagen casings (4 cm diameter) (Fibran, S.A., Gerona, Spain).
- Sausage stuffer (MAINCA, Granollers, Spain).
- Ripening chamber (temperature and humidity controlled) (Binder, model KBF 240, Tuttlingen, Germany).

Procedure (*see* **Note 15**):

1) **Meat and Fat Preparation**

(1) Select pork meat and backfat (18%) according to the specified formulation.

(2) Mince the pork meat and backfat separately using a 4.5 mm grinding plate (*see* **Note 16**).

2) **Batter Preparation**

(1) Put the minced meat (75%) and backfat (18%) in the homogenizer and homogenize them together for 2 min (Fig. 2).

(2) Add half of the total amount of additives [choravi (5.2%), NaCl (1.5%), curavi (0.3%)] and mix for 1 min.

Fig. 2 Alginate beads incorporation into chorizo batter before (**A**) and after the final homogenization (**B**)

(3) Add the remaining half of the additives along with the *L. plantarum*alginate beads (3.75%) (Fig. 2), and mix for an additional 2 min (*see* **Note 17**).

3) **Stuffing**

(1) Immediately stuff the prepared batter into collagen casings using a 4 cm diameter stuffer (Fig. 3).

(2) Tie or clip the casings as required for uniform size and shape.

4) **Ripening**

(1) Transfer the stuffed chorizos (Fig. 4) into a ripening chamber.

(2) Program the chamber with the following conditions:

- Phase 1: 23°C and 90% relative humidity for 48 .
- Phase 2: 13°C and 75% relative humidity until the end of ripening (total duration: 20 days) (*see* **Note 18**).

3 Notes

1. All steps must be conducted under aseptic conditions to prevent contamination.
2. The MRS broth should be prepared according to the producer's instructions.

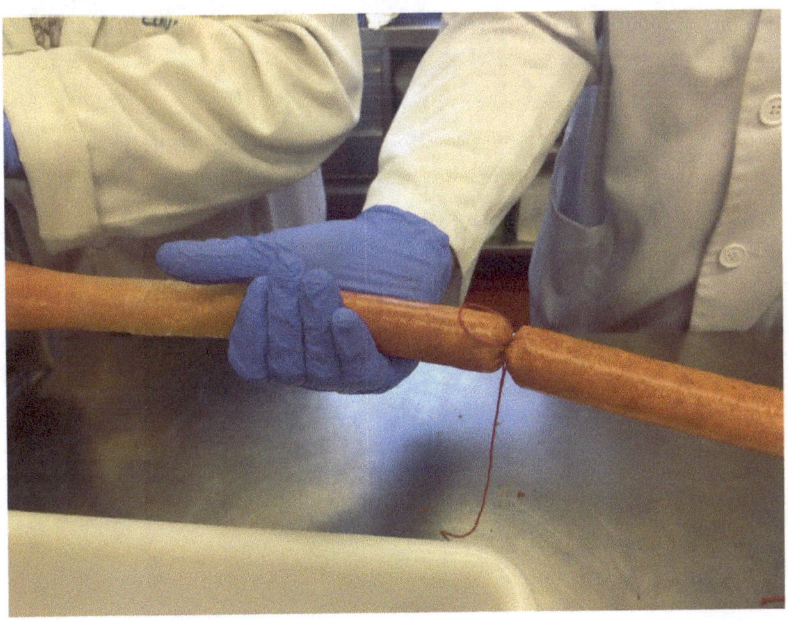

Fig. 3 Stuffing and tying processes of probiotic chorizo sausages

Fig. 4 Probiotic chorizos after stuffing (**A**) and placed in a ripening cabinet (**B**)

3. Glycerol serves as a cryoprotectant and must be mixed thoroughly before freezing.
4. Label each Eppendorf tube with the date and sample identification.
5. At the end of the procedure, a probiotic concentration of ~10^8 CFU/g will be achieved.
6. The 0.9% saline solution is prepared by mixing 0.9 g of NaCl in 100 mL of distilled water.
7. Ensure aseptic conditions throughout the procedure to avoid contamination.
8. Use an appropriate rotor and tubes for high-speed centrifugation.
9. The saline suspension should be used within a reasonable time to preserve cell viability.
10. The 0.1 M calcium chloride solution is prepared by mixing 11.1 g of calcium chloride in 1 L of distilled water.
11. All steps should be conducted in duplicate or triplicate under aseptic conditions.
12. The diameter of the pipette tip directly influences capsule size; verify calibration if necessary.
13. Ensure continuous stirring of the calcium chloride bath to prevent capsule aggregation.
14. The alginate beads are prepared according to the previous step.
15. Ensure all materials and equipment are properly cleaned and sanitized before use.
16. Maintain cold chain during meat handling to ensure microbiological safety and process control.
17. Throughout the process, ensure that the final meat batter temperature does not exceed 11°C.
18. Final product quality should be monitored by physicochemical and microbiological analyses, if applicable.

Acknowledgments

This research was supported by projects AGL2011-29644-C02-01 from MCyT; MCIN/AEI (PID2019-107542RB-C21); and CSIC Intramural Projects 201447E073, 202370E138, 202370E140, and 202570E078. This research was also funded in part by Brazilian agencies Coordenação de Aperfeiçoamento de Pessoal de Nível Superior (CAPES), Finance Code 001; CAPES-PrInt Program; and MCTIC/CNPq (Grant No. 406734/2022-4; INCT Meat Production Chain–Cadeia Produtiva da Carne).

References

1. Triki M, Herrero AM, Rodríguez-Salas L, Jiménez-Colmenero F, Ruiz-Capillas C (2013) Chilled storage characteristics of low-fat, n-3 PUFA-enriched dry fermented sausage reformulated with a healthy oil combination stabilized in a konjac matrix. Food Control 31:158–165. https://doi.org/10.1016/j.foodcont.2012.10.008
2. Prado N, Sampayo M, González P, Lombó F, Díaz J (2019) Physicochemical, sensory and microbiological characterization of Asturian Chorizo, a traditional fermented sausage manufactured in Northern Spain. Meat Sci 156:118–124. https://doi.org/10.1016/j.meatsci.2019.05.023
3. Leroy F, Aymerich T, Champomier-Vergès MC, Cocolin L, De Vuyst L, Flores M, Leroi F, Leroy S, Talon R, Vogel RF, Zagorec M (2018) Fermented meats (and the symptomatic case of the Flemish food pyramid): are we heading towards the vilification of a valuable food group. Int J Food Microbiol 274:67–70. https://doi.org/10.1016/j.ijfoodmicro.2018.02.006
4. Rodríguez-González M, Purriños L, Martínez S, Carballo J (2025) Spices affect the biochemical events taking place during the manufacture of Galician chorizo sausage. LWT 216:117322. https://doi.org/10.1016/j.lwt.2025.117322
5. Hammes WP, Knauf HJ (1994) Starters in the processing of meat products. Meat Sci 36:155–168. https://doi.org/10.1016/0309-1740(94)90039-6
6. Ruiz-Capillas C, Triki M, Herrero AM, Rodriguez-Sala L, Jiménez-Colmenero F (2012) Konjac gel as pork backfat replacer in dry fermented sausages: processing and quality characteristics. Meat Sci 92:144–150. https://doi.org/10.1016/j.meatsci.2012.04.028
7. Spaziani M, Del Torre M, Stecchini ML (2009) Changes of physicochemical, microbiological, and textural properties during ripening of Italian low-acid sausages. Proteolysis, sensory and volatile profiles. Meat Sci 81:77–85. https://doi.org/10.1016/j.meatsci.2008.06.017
8. Weinbreck F, Bodnár I, Marco ML (2010) Can encapsulation lengthen the shelf-life of probiotic bacteria in dry products. Int J Food Microbiol 136:364–367. https://doi.org/10.1016/j.ijfoodmicro.2009.11.004
9. Food and Agriculture Organization of the United Nations and World Health Organization (FAO/WHO) (2002) Guidelines for the evaluation of probiotics in food: Joint FAO/WHO Working Group Meeting Report. London, ON, Canada. https://isappscience.org/wp-content/uploads/2019/04/probiotic_guidelines.pdf
10. Klingberg TD, Budde BB (2006) The survival and persistence in the human gastrointestinal tract of five potential probiotic lactobacilli consumed as freeze-dried culture or as probiotic sausage. Int J Food Microbiol 109:157–159. https://doi.org/10.1016/j.ijfoodmicro.2006.01.014
11. Khan KI, Arshad MS, Anjum FM, Sameen A, Rehamn AU, Gill WT (2011) Meat as a functional food with special reference to probiotic sausages. Food Res Int 44:3125–3133. https://doi.org/10.1016/j.foodres.2011.07.033
12. Cavalheiro CP, Ruiz-Capillas C, Herrero AM, Jiménez-Colmenero FJ, Menezes CR, Fries LLM (2015) Application of probiotic delivery systems in meat products. Trends Food Sci Technol 46:120–131. https://doi.org/10.1016/j.tifs.2015.09.004
13. Champagne CP, Fustier P (2007) Microencapsulation for delivery of probiotics and other ingredients in functional dairy products. In: Saarela M (ed) Functional dairy products, 2nd edn. Woodhead Publishing, UK, pp 404–426
14. Pasqualin Cavalheiro C, Ruiz-Capillas C, Herrero AM, Jiménez-Colmenero F, Pintado T, Ragagnin de Menezes C, Martins Fries L (2019) Effect of different strategies of Lactobacillus plantarum incorporation in chorizo sausages. J Sci Food Agric 99:6706–6712. https://doi.org/10.1002/jsfa.9952

Index

José Manuel Lorenzo Rodriguez (ed.), *Meat Product Quality Improvement*, Methods and Protocols in Food Science, https://doi.org/10.1007/978-1-0716-5460-6,

FSC
www.fsc.org
MIX
Papier | Fördert
gute Waldnutzung
FSC® C083411

Kolibri 360 GmbH
Ferdinand-Jühlke-Straße 7
99095 Erfurt, Deutschland
produktsicherheit@kolibri360.de